TRENDS IN ZOOLOGY

TRENDS IN ZOOLOGY

Editor
CHARULATA SHARMA

KALPAZ PUBLICATIONS

DELHI-110052

TRENDS IN ZOOLOGY

(Zoology)

ISBN: 81-7835-048-3

Published in 2001 in India by
Kalpaz Publications
C-30, Satyawati Nagar
Ashok Vihar Phase-III
Delhi-110052
E-Mail: kalpaz@hotmail.com

Laser Typesetting at : A.S. Printers
Printed at : Mehra Offset Printers, New Delhi

Preface

Zoology can be defined fairly easily by saying that it is the science of animal life. As one looks at the vast array of animal kingdom, one is impressed with their tremendous diversity. With all the diversity, there is a certain unity, a certain paucity of basic patterns of animals in the world. The book covers virtually all aspects of zoology in a concise and well illustrated manner.

This book presents fundamental aspects of zoology within the framework of scientific discovery. Researchers have made almost incredible strides in the past decades. Consequently existing trends in zoology have expanded. The text integrate the descriptive, experimental and biochemical approaches into a conceptual framework, for the analysis of developments in the subject.

The book consists of up-to-date accounts on invertebrates, vertebrates, cytology, genetics and evolution. The first chapter presents a broad survey of invertebrates with descriptive taxonomy, structure, reproductive processes, economic importance, and phylogenetic relationships. The second chapter briefly describes cellular and extracellular environment of organisms: The chapter on genetics describes the fundamental aspects of genes and heredity. The fourth chapter provides a systematic study of vertebrate physiology. The chapter on evolution provides a step-by-step analysis of evolutionary origin of animals. The increased depth of coverage in this chapter was prompted by the recent revival of interest in evolutionary biology.

The book contains informative and comprehensive account on the subject in light of new findings. It is written for introductory coursed in zoology. Students and teachers in field of biological science find this book highly useful.

Editor

Contents

Chapter 1

World of Invertebrates

Protozoa

Protozoans are singlecelled organisms possessing typical membrane-bound cellular organelles. Because many are animal-like, being motile and having heterotrophic nutrition, this assemblage was treated in the past as a single phylum within the Animal Kingdom—the phylum Protozoa. Protozoa exhibit all types of symmetry, a great range of structural complexity, and adaptations for all types of environmental conditions.

Although most protozoa occur as solitary individuals, there are numerous colonial forms. Some of these, such as species of *Volvox*, attain such a degree of cellular interdependence that they approach a multicellular tissue level of structure. Individual independent protozoa range in size from a micrometer in the case of the planktonic Micromonas to a few millimeters in some dinoflagellates, amoebas, and ciliates.

Free living protozoa occur wherever moisture is present in the sea, in all types of fresh water, and in the soil. There are also commensal, mutualistic, and many parasitic species. The protozoan body is usually bounded only by the cell membrane. The rigidity or flexibility of the body and its shape are largely dependent on the nature of the cytoskeleton. Cytoskeletons are typically but not always located immediately below the cell membrane, where, together with it and, other organelles, they form the *pellicle*, a sort of protozoan "body wall."

The cytoskeleton is often composed of slender filamentous proteins, microtubules or vesicles, or various combinations of all three. The filamentous proteins may form a dense supportive mesh called the epiplasm, as, for example, in Euglena and the ciliates. More conspicuous cytoskeletal of ganelles are the pellicular microtubules that occur in all flagellates, in spore-forming protozoa, and in ciliates. They can be arranged as a microtubular corset, of, as in some flagellates, the microtubules originate on the flagellar basal bodies and radiate rearward from them to the opposite extremity of the cell as a sort of axial skeleton (axostyle). Such microtubules resemble the microtubules of a mitotic spindle, which radiate between centrioles and form the mitotic apparatus.

In other protozoans, such as the radiolarians and heliozoans, bundles of microtubules radiate outward from a centrosome or similar structures (both MTOCs), each bundle extending into and supporting a raylike projection of the cell (axopod). The centrosome and its microtubules resemble the starlike asters that form around centrioles at the poles of the mitotic spindle. Vesicles occur immediately below the cell membrane in many protozoans, such its dinoflagellates spore- forming protozoans, and ciliates. When these vesicles are flattened and form a more or less continuous layer beneath the cell membrane, they are referred to as alveoli. "Empty" alveoli, like those that occur in ciliates, may be turgid and help to support the cell. In dinoflagellates, plates of cellulose are secreted into the alveolar vesicles to form a rigid skeleton.

Protozoan skeletons, like those of metazoans can also be exoskeletons. They are secreted onto the outer surface of the cell and are usually called tests.The protozoan locomotor organelles may be flagella, cilia, or flowing extensions of the body called pseudopodia. These are important organelles for defining the protozoan groups and will be discussed extensively in later sections of the chapter.

All types of nutrition occur in protozoa. Some protozoa rely on photoautotrophy, others absorb soluble organic nutrients from their environment, and many ingest food particles or prey and digest them in tracellularly within food vacuoles. Like other organelles, the walls of food vacuoles are composed of lipid bilayer membranes, which can be synthesized or removed rapidly and recycled. Food reaches the vacuole by phagocytosis, often through a cell mouth or cytostome. Very fine particles in solution may enter by pinocytosis, which may occur over the entire surface of the cell. Intracellular digestion has been most studied in amoebas and ciliates, and for the most part, follows the general pattern.

Protozoans live in the digestive tract of animals, are obligatory anaerobes. Those that live in water where there is active decomposition of organic matter may be facultative anaerobes, utilizing oxygen when present but also capable of anaerobic respiration. The changing availability of food and oxygen associated with decay typically results in a distinct succession of populations and protozoan species.

Because of their short generation time, protozoa can be important in monitoring pollution. Their relatively rapid response to changes in the aquatic environment can be indicative of reduced oxygen. Many protozoa osmoregulate to remove excess water and to adjust the concentration and proportion of their internal ions. Excess water enters protozoans by osmosis when their internal osmotic concentration exceeds that of the surrounding water. Additional water may enter with food in vacuoles and pinocytotic vesicles, for example, an amoeba that is experimentally fed on a protein solution imbibes by endocytosis a quantity of water equivalent to one third its body volume. The complex is composed of a large spherical vesicle—the contractile vacuole proper—and a surrounding system of small vesicles or tubules termed the spongiome. The spongiome provides for the collection of fluid, which is delivered to the contractile vacuole, which in turn expels the fluid to the outside of the organism through a temporary or permanent pore.

In some protozoa (some amoebas and flagellates), the vacuole completely disappears following contraction and is reformed by fusion of small vesicles. In others (many ciliates), the vacuole collapses at discharge and is refilled by fluid from the surrounding tubules of the spongiome. The rate of discharge depends on the osmotic pressure of the external medium. Like animals, protozoa respond to chemical and physical cues in ways that enable them to avoid adverse conditions, locate food, and find mates.

In this sense, each protozoan cell must be both receptor and effector. In their receptiveness to environmental stimuli, protozoans resemble the sensory nerve cells of animals. Like animal receptors, protozoans often receive external stimuli as signal substances that bind to specific membrane molecules.

Asexual reproduction by mitosis occurs in most protozoa and is the only known mode of reproduction in some species. Division of the organism into two more progeny cells is called fission. When this process results in two similar progeny cells, it is termed binary fission; when one progeny cell is much smaller than the other, the process is called budding. In some protozoa, multiple fission, or schizogony, is the rule. In schizogony, after a varying number of nuclear divisions, the cell divides into a number of progeny cells. With few exceptions, asexual reproduction involves some replication of organelles following fission. Sexual reproduction is not universal among protozoans and where it occurs, commonly does not lead directly to the formation of new individuals.

Coelenterata

Coelentrates are radially symmetrical, usually with tentacles; body of two cellular layers with a jellylike layer between; with only one cavity in the body, a digestive cavity, the enteron, having only one regular opening to the outside the mouth. Almost all animals are aquatic, mostly marine and none parasitic. Hydroid forms generally cylindrical, often in colonies with tentacles at free end; medusoid forms umbrellalike, with tentacles around margin; other body layer (epidermis) composed of a number of types of including specialized stinging cells, the nematocysts, which are used to stun or capture prey; inner body layer (gastrodermis) similar in cell types, but with digestive cells; between these is jellylike mesoglea, thin in hydroids and thick in medusoids. Body essentially two layers of integument, with complex specialization of cells.

Mostly hydrostatic skeletons, if any; corals lay down proteinaceous or calcareous skeletons of various sizes and complexities. Epithelial cells of two body layers have muscle fibrils at the bases, extending generally around and lengthwise of the body; these may become separate in jellyfish and form more complex patterns; hydroids mostly sessile; medusoids float or swim by contracting edges of the umbrella; tiny larvae swim by cilia.

Food mostly animal matter, often captured by tentacles; ingestion through mouth into the blind cavity, the enteron; digestion usually in cells lining the enteron following preliminary disruption of body of prey; foods-move from place of digestion to other cells by diffusion;

egestion through mouth. There was not specific respiratory organs. Each cell acts separately. Regulatory and excretory is also performed by each cells. Nervous system is netlike, mostly in the epidermis; conduction generally nonpolarized, no specialized ganglia-reactions generally gross and general. Both asexual and sexual reproduction are taking place in animals coming under this phylum.

Platyhelminthes

Platyhelminthes are aquatic: marine, freshwater, or parasitic; some live in moist soil, etc. (land planarians), and many are parasitic, living in blood, intestinal contents, liver, etc. Bilaterally symmetrical i.e., with body divisible into right and left halves by a median plane, flattened worms, with head variously developed but may simple: either solid or with only one cavity in the body—the digestive cavity or enteron—which has only one opening to the outside. Body flattened and very thin, usually with three cellular layers:

i) Epidermis, usually thin and one cell layer thick.

ii) Gastrodermis, lining the enteron and

iii) Parenchyma between epidermis or gastrodermis, variably developed, usually complex and containing the internal organs.

Movements brought about by muscles in the parenchyma; muscles mostly extending across and around the body, but some also vertical and diagonal. Locomotion: cilia on epidermis in larvae and adults of many; muscular contractions aided by adhesive structures in others. Food generally animal matter; two groups are parasitic; ingestion through mouth or by absorption of already digested materials through body wall; digestion, when necessary, mostly in the gastrodermal cells; foods diffuse to other parts of body;

egestion, where necessary, through mouth. No specialized respiratory system; each cell acts separately. Regulation and excretion are by flame cells and excretory tubes present, which are probably mostly for water and ion regulation: excretion generally through body wall.

Nervous coordination system is well developed. Reception of stimuli by general sensitivity over body; parasites seem relatively insensitive, or at least unresponsive; specialized receptors present in, some species. e.g., eyespots, current receptors on auricles, statocysts etc. Nervous system well developed, with a centralized anterior ganglion and nerve cords and peripheral nerve-net wider the epidermis; conduction polarized; reactions generally gross, usually avoidance and approach or shortening and lengthening.

Both asexual and sexual reproduction are taking place in animals coming under this phylum. Asexual reproduction ocuurs by all forms widespread; regenerative powers very great; budding produces segment like body of tapeworms. Sexual reproduction is generally complicated reproductive system particularly elaborate in parasites, which produce prodigious quantities of eggs: all the ducts, glands, etc., listed as secondary sexual structures are present in most species, along with other specialized structures: mostly hermaphroditic, but some bisexual.

Nematoda

Nematodes fundamentally aquatic, but many are parasites in plants and animals and many live in moist soil, plant sap, decaying vegetation, and animal matter. The animals are bilaterally symmetrical, usually wormlike, slender, and almost cylindrical; with an unlined space between the body wall and digestive tract; the latter open at both ends of the body, i.e., mouth and anus present. Body usually slender, cylindrical and tapered at each end; generally little or no external ornamentation or projections; body has three cellular layers: i) outer epidermis which secretes a tough, pliable cuticle; ii) muscle layer. No special respiration system; may be through body wall or intestinal wall, but mostly unknown; some apparently do not need oxygen. Regulation and excretion by special cells, called renettes, and excretory canals are present in many; little is known about the function of these.

Epithelia generally one cell layer thick and simple in appearance; epidermis secretes cuticle; no cilia occur on any cells. Skeletal support is done by which forms tough, resistant but resilient exoskeleton, acting against internal pressure to give a pliable hydrostatic skeleton. Movements brought about by muscles pulling on body wall creating undulations; little shortening or expansion possible because of cuticle; locomotion effected by swimming undulations. Food varies greatly among different species; free-living species take in all kinds of organic matter, both plant and animal; many species are parasitic on plants and animals; ingestion is through mouth aided by a sucking pharynx and in some cases toothlike structures; digestion in intestine, probably by enzymes secreted by cells of the intestinal wall, but many species ingest already digested materials; absorption into pseudocoel; egestion through anus.Nervous system consists of ringlike ganglion at the front end with two or more nerve cords extending along the body; and a nerve net in the body wall. Behaviour patterns generally simple avoidance and approach.

Generally, sexual reproduction is taking place. The reproductive system is complex, particularly in parasites, including all the ducts, glands, etc. and in some cases other special structures; mostly bisexual; eggs covered with shells, etc., either deposited in a cocoon, released to the outside, or held until young hatch. Life history usually hatching young is like adult in appearance, but may he larval in respect to its differences in behaviour or habitat; free-living species generally have simple development; parasites usually have rather direct developmental patterns that may involve more than one host, but some have complex life cycles.

Annelida

Animals coming under this phylum are generally aquatic: marine, freshwater, or moist soil, water-filled leaves, et mostly free-living, but many leeches are parasites. Animals are wormlike, bilaterally symmetrical, with the body divided into segments externally and internally; the body wall generally has bristlelike or hairlike setae; with a coelom; all organ systems, excluding respiratory, well developed.

Annelidens have wormlike like body, i.e., more or less slender and cylindrical, but some are flattened and some have projections (parapodia) from the sides of the body; the body wall is relatively complex, with an epidermis that secretes a cuticle, variously developed, at least two muscle layers: longitudinal mid circular, and an inner epithelium lining the body cavity which is thus a true coelom; the digestive tube has corresponding layers; the body is divided into segments externally and internally, each segment generally traversed by the digestive tube and nervous system, may contain other segmentally repeated internal organs; all systems, except respiratory, are usually well organized.

Generally delicate, one cell layer thick; epidermis usually producing a cuticle, which may be very thin and soft (earthworms) or have bristles and scales. Usually only a hydrostatic skeleton is present, produced by contractions of muscles of body wall acting on coelomic fluid; many marine worms produce tubes of various permanence and rigidity. Movements brought about by muscles of body wall acting as opposing sets resulting in peristaltic waves; many also have diagonal muscles and muscles in the parapodia, where present, that produce special movements. Locomotion in adults usually by undulating swimming or in terrestrial species by crawling, using bristlelike setae in body wall for attachment; burrowing in bottom mud or soil is performed by driving forth either the body or pharynx as a hydraulic ram; larvae and some adults of aquatic species swim by cilia on epidermal cells.

Foods are widely various, almost any organic matter; mostly free-living but one group includes many parasites which suck blood; ingestion through mouth aided by suction of muscular pharynx, in some the mouth has cutting or piercing structures; digestion in complete and mainly complex digestive tract, consisting of the usual parts; digestion occurs in the intestinal cavity, and foods are absorbed into the blood; egestion through anus. Regulation and excretion are generally by one pair of nephridia in most segments of the body; nephridia in aquatic species mostly regulate water and ions, excretion being through body wall; in terrestrial species nephridia are both regulatory and excretory.

Nervous system well developed, with a central ganglion above the pharynx, connected by cords around the pharynx to a nerve cord along the lower side, generally with a ganglion in each segment from which peripheral nerves lead to the sense organs and muscles, etc. Reactions and behaviour: generally gross, whole body responses to stimuli, but capable of some complex behaviour, e.g., tube construction, mating behaviour, food selection, etc.

Both asexual and sexual reproduction are taking place in animals coming under this phylum. Asexual reproduction is by budding and fission in many species;

regenerative powers generally excellent, but selective (e.g., some parts regenerate easily, others not so). Methods of sexual reproduction varies from species to species; some marine worms have only gonads, the eggs and sperms leaving the body either through nephridia or by rupture of the body wall; earthworms have complex systems with the usual secondary sexual structures of such systems, many (e.g., marine worms) are bisexual, others (e.g., earthworms) are hermaphroditic; eggs usually develop in some protective case outside body.

Echinoderamata

Adults generally are radially symmetrical, with star-shaped patterns, or a radially arranged cylindrical body form; without head or segmentation; with tube feet and an endoskeleton of calcareous ossicles; all marine.

Body of adults generally built on radially repeated patterns of five sectors; usually with plates forming spiny skeleton or boxlike case; one class (Holothunioidea) has a five-part radial pattern, but the body is cylindrical and skeletal elements often scattered and small; no definite head present; body instead divisible into oral surface, with mouth, and aboral surface.

Epidermis generally single cell thick and delicate, with mucous glands; often brightly coloured; dermis variably developed and producing skeletal elements. Endoskeletal ossicles laid down by dermis are variable in pattern: may form platelike structures, often with spines projecting through epidermis, may fuse into solid, boxlike case, or may be scattered about loosely in dermis. Movements by action of muscles in bands pulling on ossicles; in sea urchins, skeleton prevents movements; in others whole-body movements occur.

Locomotion mostly by action of tube feet which are hydraulically operated by seawater in a system of tubes, the water-vascular system, opening through a pore, the madreporite, to the ocean; this peculiar system is characteristic of the echinoderms.

Food various; generally carnivorous or scavengers, capturing food with tube feet or tentacles and moving it to mouth; ingestion through mouth, which may be unarmed or, in sea urchins, have a very complex scooplike set of ossicles, called Aristotle's lantern; digestion in stomach and in caeca opening into stomach; absorption from stomach, caeca, and intestine; stomach and intestine variously developed in group; brittle-stars have neither intestine nor anus; egestion generally through anus, except in brittle-stars.

Neuroendocrine cells present in nervous system; generally almost nothing known for sure. Nerve ring around mouth, with nerve cords extending radially along canals of water-vascular system; in general, little known about nervous action. Reactions and behaviour very simple behaviour patterns and capabilities.

Sexual reproduction is by simple systems, only gonads and short ducts to outside; external fertilisation, with eggs either released free in ocean.

Mollusca

Molluscans are bilaterally symmetrical animals, but the body is often twisted or spiralled, without segmentation: body generally consists of three major regions: i) bead, usually a sensory center; ii) visceral mass, containing highly developed internal organs; and iii) foot, the locomotor organs generally enclosed in a shell secreted by the outer covering of the body, the mantle. Head region variously developed, level of develo-pment related chiefly to habits of animal-free-living active species e.g., land snails, octopus) have well-developed head region with specialized sense organs; sessile or burrowing species (e.g., oyster, clams) have little or no externally visible head; foot is strongly muscular, adapted for craving (e.g., snails), swimming or grasping (e.g., octopus), or burrowing (e.g.. clam); visceral mass varies in shape and size, often twisted in shell (e.g., snails),

Generally, one cell layer epidermis, very delicate; may be ciliated, or in some have dense mucous covering or a cuticle; epidermis of mantle highly specialized for shell production and repair. Generally no internal skeleton, except for those having the shell beneath the mantle rather than outside, shells secreted by mantle are characteristic of the phylum, and these form the major skeletal element, although hydrostatic skeletons, particularly in the foot, are important. Cilia are widespread in phylum, but function for locomotion almost exclusively in larvae; muscles of a number of types, still not clearly separated, are the usual kinetic structures, occurring in complex patterns of bands and bundles rather than circularly and longitudinally only.

Movements are generally fairly gross and restricted, except for tentacles and other projections of body wall. Locomotor patterns widely various, brought about mainly by muscular actions of foot, or rarely the mantle: gliding locomotion by waves of muscular contraction in the foot of snails and slugs, with points of attachment; digging by extension and retraction of bladelike foot of clam, with anchorage of the tip by hydrostatic expansion; swimming ciliary action in larvae or tiny molluscs, or by movements of mantle flaps in some sea slugs, or by squirting of water from the mantle cavity through a jet in squid and octopus. Foods include almost any organic materials-algae and other green plants, bacteria, other animals, wood, etc.; food collection is extremely varied e.g., filter feeding in clams, use of tentacles in octopus, grazing and predation in snails, etc.; ingestion through mouth, which is often armed with a special rasping structure, the radula; digestion in a complete and complex digestive tract, including esophagus, stomach, digestive gland, intestine and anus; digestion variously in stomach or intestine or in amoebocytes which enter digestive tract and carry food particles out to other parts of body, absorption of food from digestive gland and intestine into the blood; egestion through anus.

Closed circulatory system, but with large and very well-developed blood sinuses; pumping organ is generally a well-developed heart, often with one or more auricles and a ventricle; in many species accessory hearts are also present near gills, etc.; blood vessel patterns differ widely in phylum: generally main arteries go to head, foot, visceral hump, and mantle, where sinuses allow blood to trickle through in close proximity to the tissues, from the sinuses blood returns in veins to heart through gills and kidneys where wastes are lost and oxygen is taken in. Neuroendocrine cells in the nervous system and some other parts are known, but their functions are almost entirely unknown. Nervous system very well developed and generally centralized; consists basically of three major ganglia: i) cerebral ganglion in the head, connected by cords with the ii) pedal ganglion in the foot and iii) visceral ganglion in the visceral mass; ganglia act generally as local reflex centers, with some central control from brain; in more highly specialized molluses the ganglia are all in the head fused into a true central brain.

Both asexual and sexual reproduction are taking place. Sexual reproduction is tremendously varied in the group; sessile species, such as clams and oysters, generally have only gonads and short ducts through which sperms and eggs are discharged to the outside; terrestrial snails and many marine snails have highly complex systems, involving the usual array of secondary sexual structures, along with many special structures occurring only in those animals; molluscs are usually oviparous, producing eggs with membranes, shells and capsules or other protective coverings; some are viviparous, mostly bisexual.

Arthropoda

Arthropods are the most successful animals today, their members invading all habitats; even though arthropods do not have constant body temperatures, they have been found everywhere on earth and at very high altitudes in air; they range in habits from the most highly specialized parasites known to social forms such as bees, termites, and ants. Bilaterally symmetrical animals with segmented body, usually having three regions: head, thorax and abdomen; covered by an exoskeleton which is periodically molted, and with jointed appendages.

A vast and varied group (about 700,000 species); body usually has head, thorax, and abdomen, but head and thorax may be joined (e.g., crayfish, spider), or thorax and abdomen similar (e.g.. millipedes); head is generally sensory area often with antennae, eyes, etc.: thorax is generally locomotory, with legs, wings etc.; abdomen generally, houses reproductive organs, another viscera, and has external reproductive appendages; appendages are jointed and of many types: antennae, mouthparts, leg, swimmerets, claws, genitalia, etc.

Epidermis is single cell layer, varying in thickness at various stages exoskeletal production; glandular elements produce numerous secretions that are part of or cover exoskeleton. Exoskeleton, including that of appendages, secreted by epidermal cells, is characteristic of phylum; consists of tanned proteins, chitin, and in some calcium carbonate, in variable proportions producing soft to very bard covering to body; exoskeleton shed, or molted, periodically to allow growth, the molting process initiated and controlled by hormones. Movement is generally regional and precise through bending at soft places (joints) in body or appendages; locomotion widely varied-swimming, creeping, burrowing, running, flying, etc—generally results from movements of appendages activated by muscles; in many adult insects, large muscle bands in the thorax pull on the walls to raise and lower the wings rapidly, while small muscle strands turn and tilt wings to give propulsive energy.

Food various, depending upon species—all types of organic matter taken by some, others restricted to only one species of plant or animal for food; scavenging, predation, grazing, etc., common; methods for securing food and feeding equally varied-filter feeding, grasping and chewing (e.g., crab, grasshopper), sucking (e.g., spider, butterfly), etc.; ingestion is through mouth which is simple, but finished with special appendages, the mouthparts, that aid in chewing or sucking; digestion in a complete and often very complex digestive tract, with esophagus, stomach, intestine, and anus; digestion and absorption in cavity of stomach, intestine, or caeca egestion through anus.

Aquatic species generally have thin-walled projections of the body wall near the bases of the legs, the gills, over which water is moved by gill bailers; some respiration occurs also through thin areas between segments; terrestrial species have either book lungs at the base of the abdomen e.g., spiders. Regulatory and excretory structures, basically like nephridia, acting in both regulation mid excretion: green glands at bases of antennae in crustaceans, coxal glands at bases of legs in scorpions, etc..

Malpighian tubules emptying into the digestive tract, functioning as both the regulatory and excretory organ in insects and some other terrestrial arthropods. Distributive system is an open type with blood free in capacious spaces (hemocoel) in body and appendages; in aquatic species blood is generally drawn from the hemocoel into veins which conduct it to gill capillaries, whence it goes in short veins to pericardium and thence into the heart, which pumps it in open-ended arteries to various parts of the body, and thus into the hemocoel again; in terrestrial species, the veins are small or lacking, there are no gills, and the arteries are similarly short; in higher insects, arteries and veins are almost nonexistent, and the heart, within a pericardium, merely stirs the blood in the hemocoel; in some arthropods blood has respiratory pigments—e.g.. hemocyanin, a blue pigment, in some crustaceans; blood usually lacks respiratory pigments in insects.

Neuroendocrine cells present in the nervous system in many places; endocrine glands present-e.g., X organ and Y organ of crustaceans, corpora allata, and prothoracic glands of insects; molting generally under endocrine control; various aspects of metabolism and reproduction also known to be coordinated by neurohormones and endocrine hormones.

Nervous system segmentally arranged, somewhat like annelids, with central ganglion in the head above the esophagus (supraesophageal ganglion or brain) connected by cords around the esophagus (circumesophageal commissures) to a ventral nerve cord, which usually has a ganglion in each segment; in higher insects these ganglia gather forward to form a highly central mass with the subesophageal ganglion.

Reproductive systems vary from extreme simplicity (e.g., in a crustacean, Sacculina, that is parasitic in crabs, the mature individual is a mere bag of eggs or sperms which are released to the ocean by rupture of the body) to almost improbable complexity, with the usual secondary sexual structures, many more glands, etc.; in insects, transfer of sperms and deposition of eggs involves specialized abdominal appendages, the genitalia, that may be very complicated, often forming lock-and-key-like arrangements in male and female that are characteristic of each species usually these sexes are separate; oviparous with egg coverings or brooding., or viviparous: in many species elaborate courtship, mating, and parental behaviour.

Importance of insects

The class Insecta, or Hexapoda pf phylum arthropoda containing more than 750,000 described species, is the largest group of animals; in fact, it is three times larger than all the other animal groups combined. Only a brief, rather superficial treatment of insects is possible here.

Insects are distinguished from other arthropods by having three pairs of legs and usually two pairs of wings carried on the middle, or thoracic, region of the body. In addition, the head typically bears a single pair of antennae and a pair of compound eyes. A tracheal system provides for gas exchange, and the gonoducts open at the posterior end of the abdomen.

The success of insects is evidenced by the tremendous number of species and individuals and by their great adaptive radiation. Although they are essentially terrestrial animals and have occupied virtually every environmental niche on land, insects have also invaded the aquatic habitats and are absent only from the subtidal waters of the sea. This success of insects can be attributed to a number of factors, but certainly the evolution of flight endowed these animals with a distinct advantage over other terrestrial invertebrates. Dispersal, escape from predators, and access to food or optimum environmental conditions were all greatly enhanced. The powers of fight also evolved in reptiles, birds, and mammals, but the first flying animals were insects. Insects are of great ecological significance in the terrestrial environment. Two thirds of all flowering plants depend on insects for pollination.

Insects are also of enormous importance for humans. Mosquitoes, lice, fleas, bedbugs, and a host of flies' can contribute directly to human misery. More importantly, these and others affect us- indirectly as vectors of human diseases or of diseases of domesticated animals: mosquitoes (malaria, elephantiasis, and yellow 1 ever) tsetse fly (sleeping sickness); lice (typhus and relapsing fever): fleas (bubonic plague): and the housefly (typhoid fever and dysentery).

Our domesticated plants are dependent on some insects for pollination but are destroyed by others. Vast sums are expended to control insect pests, which can greatly reduce the high agricultural yields necessary to support large human populations. The overzealous use of pesticides, however, can in turn be hazardous to the environment and to human health.

Although the exoskeleton of an arthropod segment is composed of a tergum, a sternum, and two pleura, some parts are more highly sclerotized or are more conspicuous than others. Such thickened areas of cuticle, called sclerites, are prominent features of the insect body surface. They are often separated by sutures, which reflect ridges on the inner side of the skeleton or lines of thinner exoskeleton.

The heads of most insects are oriented so that the mouthparts are directed downward. A more specialized, anteriorly directed position (prognathus) is found in some predaceous species, such as the carabid and tiger beetles; a posteriorly directed position (opisthognathus) is present in the hemipterans and homopterans, which have sucking beaks. The head skeleton of most insects forms a complete, external capsule surrounding the soft, inner tissues.

The more lateral and dorsal surfaces of the head bear one pair of compound eyes and one pair of antennae. Between the eyes and the antennae are usually three ocelli. Three pairs of appendages contribute to the mouthparts. One pair of mandibles is located anteriorly, followed by a pair of maxillae and then by the labium. Although single, the labium actually represents a fused pair of second maxillae.

Anteriorly, the mandible is covered by a shelflike extension of the head, forming an upper lip, or labrum. From the floor of the prebuccal cavity projects a median lobelike process, called the hypopharynx. The hypopharynx arises behind the maxillae near the base of the labium. The modifications of the mouthparts associated with different feeding habits will be discussed later.

The thorax, which forms the middle region of the insect body, is composed of three segments: a prothorax, a mesothorax, and a metathorax. A pair of legs articulates with the pleura on each of the three segments. The thoracic terga of insects are called nota, and it is with the nota, and it was processes of the mesothorax and the metathorax that the two pairs of wings articulate. The basal section of the leg articulating with sclerites in the pleural area is the coxa, which is followed by a short trochanter. The remaining sections consist of a femur, a tibia, a tarsus, and a pretarsus. The tarsus is composed of one to five articles. The pretarsus is represented chiefly by a pair of claws.

In other arthropods, such as arachnids, the pretarsus is not usually considered separate from the tarsus. The legs of insects are generally adapted for walking or running, and during their effective strokes, the forelegs pull while the middle and hind legs push. The middle legs step outside of the other two pairs, which reduces interference.

One or more pairs may be modified for such functions as grasping prey, jumping, swimming, and digging.

The abdomen is composed of 9 to 11 segments plus a telson, but the telson is complete only in the primitive proturans and in embryos. The only abdominal appendages in the adult are a terminal pair of sensory cerci borne on the eleventh segment. The reproductive structures are thought by some entomologists to represent segmental appendages, but except for the female ovipositors, this is by no means certain. A variety of abdominal appendages serving different function's are present in many insect larvae.

Wings are characteristic features of insects, but a wingless condition occurs in a number of groups. In some the absence of wings is obviously secondary. For example, ants and termites have wings only at certain periods of the life cycle, workers always lack wings. Some parasitic insect orders, such as the lice and fleas, have lost the wings completely.

On the other hand, it is fairly certain that the insects in the orders Protura, Thysanura, Collernbola, and a few others arose from ancestral wing-less insects. Those groups in which the wingless condition is considered primary are called apterygotes and are believed to represent the most primitive members of the class. Winged insects and those that are secondarily wingless are referred to as pterygotes. Some entomologists consider the wingless apterygotes to be related to insects but not insects themselves. The sequence of events in the evolutionary development of insect wings is unknown.

The earliest known fossil insect is a bristletail, a wingless insect, from the early Devonian period. Winged insects appear later. No intermediate types have yet been discovered. The most widely accepted theory of wing origin is that wings were originally flat, lateral flanges of the notum, which enabled the insect to alight right side up when jumping. These flanges then gradually enlarged into winglike structures, making gliding possible. The last step was the development of hinges, enabling wings to move.

The earliest functional wing was thought to be a fan-shaped membrane with trusslike supporting veins. Because wings are evaginations, or folds, of the integument, they are composed of two sheets cuticle. The two cuticular membranes are separated by tubular thickenings called veins, which form effective supporting skeletal rods for the wing. Wing veins, open into the body and contain circulating blood and commonly tracheae and sensory nerve branches. The blood is important in maintaining the proper water content of the cuticle.

The wing venation of the more primitive insects is netlike, but there has been a general tendency in the evolution of wings toward reduction to a few longitudinal veins and cross veins, thus giving a stronger support system to the wing. The arrangement of veins in a wing is very specific in certain genera and families of insects and provides a useful tool for systematists; the principal veins and their branches are all named and numbered. Primitively, wings are held outstretched, as in dragonflies. The evolution of sclerites in the wing base, which permits many insects to fold the wings over the abdomen and thus keep them out of the way when at rest, was an important event in the evolution of the class.

The ancestors of many modem orders were then able to radiate into microhabitats, beneath bark and stones and in soil, dung, and wood, where outstretched wings would have been a serious impediment. Wing folding was probably accompanied by reduction in the body size of many groups. There is great variation in the wings of insects.

Many of these variations represent modifications that accommodate, the demands of flight characteristic to the particular group of insects. Primitively, as in damselflies, roaches, and termites, the two pairs of wings beat independently of each other, but this requires that the hind wings operate in the air turbulence created by the forewings.

Thus, in many insects the two wings on each side arc coupled by interlocking devices or by simple overlapping so that the wings operate together. Only the second pair of wings is used for flight in beetles the front pair has been adapted as hard protective plates, called elytra. Each wing articulates with the edgc of the tergum, but its inner end rests on a dorsal pleural process, which acts as a fulcrum. The wing is thus analogous to a seesaw off center. Upward movement of the wing results indirectly from the contraction of vertical muscles within the thorax, depressing the tergum.

Downward movement of the wings is produced directly, by contraction of muscles attached to the wing base (dragonflies and roaches), indirectly by the contraction of longitudinal muscles raising the ' tergum (bees, wasps, and flies), or by both direct and indirect muscles (grasshoppers and beetles). Up and down movement alone is not sufficient for flight. The wings must at the same time be moved forward and backward. The wings of many insects describe a "clap and peel" motion in the beat cycle. At the top of the upstroke the wings touch, or clap, and then peel away from each other beginning with the front margins. Air rushes between them and forms a vortex around each wing. In addition to the angle at which the wing is held, insects can also achieve lift by raising and lowering wing veins, thereby changing the wing shape, or contour. Some insects can reduce the wing surface area by changing the degree-of fore- and hindwing overlap.

All of these changes-angle, contour, and surface area—can take place in the course of one wing beat cycle and increase net upward lift. They reflect, in part, the ability of the wing to deform, rather like a sail. The raising or lowering of the wings resulting from the contraction of one set of flight muscles stretches the antagonistic muscles, which then also contract.

Insect wing beat thus involves the alternate contraction of these antagonistic elastic systems. The beat frequency varies greatly—4 to 20 bps (beats per second) in butterflies and grasshoppers; 190 bps in the honeybee and housefly; and 1000 bps in certain midges. At low frequencies (30 bps or fewer) there is usually one nerve impulse to one muscle contraction. At higher frequencies, however, the contraction is myogenic, originating from the stretching caused by the contraction of the antagonistic muscles, and there are a number of beats, or oscillations, associated with each nerve impulse. A very slight decrease in muscle length during contraction can bring about a large movement of the wing.

The elastic nature of the thoracic skeleton and the joints of wing articulation also contribute to the beat motion. If, for example, the stable position is down, as in grasshoppers, elastic forces to return to the down position are set up when the wings are raised, much like stretching a rubber band. Some flies and beetles have both stable down and stable up positions, and at some point in the movement of the winfrom one position to the other, the opposite elastic forces take over. This arrangement is called a click mechanism. Flying ability varies greatly. Many butterflies and damselflies have a relatively slow wing beat and limited maneuverability.

At the other extreme, some flies, bees, and moths can hover and dart. From the standpoint of manueverability, a housefly can outperform any bird. Not only can a housefly fly a rapid straight course and hover, it can fly upside down and turn in the distance of one body length. The fastest flying insects are hummingbird moths and botflies, which have been clocked at 25 mph. Honeybees can cruise at 15 mph. Gliding, an important form of flight in birds, occurs in only a few large insects. There is no single flight control center in the insect nervous system, but the eyes and sensory receptors on the antennae, head, wings and other parts of the body provide continuous feedback information for flight control. Horizontal stability is maintained in pail by a dorsal light reaction: the insect keeps the dorsal ommatidia of the eyes under maximum illumination from above.

Deviation because of rolling is corrected by slight changes in wing position to bring the dorsal part of the eyes back to maximum illumination. Members of the order Diptera (flies, gnats, and mosquitoes) have the second pair of wings reduced to knobs, called halteres. The halteres beat with the same frequency as the forewings and function as gyroscopes to offset flight instability. Receptors on the haltere base detect deviating forces, such as tendencies to pitch, roll, and yaw, and from this information corrections in wing position are made. Flight speed is probably determined by air flow over receptors on the antennae and movement of objects from front to back across the eyes. Flight is inhibited by contact of the tarsi with a solid surface.

Insect flight muscles are very powerful. The fibrils are relatively large, and the mitochondria are huge (about half the size of a human red blood cell), reflecting the high respiratory rate of these cells. Insects are the only ectothermic fliers, and a low body temperature and a correspondingly low metabolic rate impose limitations on mobility. On a cool morning many flying insects literally warm up before flight.

Insects have adapted to all types of diets. The mouthparts may be highly modified, but the modifications are associated less with diet than with the method by which food is obtained. Primitive mouthparts are adapted for chewing, and it is mouthparts of this type that are described and illustrated in the beginning section on external structure.

The mandibles are heavy and capable of cutting, tearing, and crushing, and the maxillae and labium function in food handling. The hypopharynx aids in swallowing. Insects with chewing mouthparts include the primitive apterygotes, dragonflies, crickets, grasshoppers, beetles, and many others. The larvae of such insects as moths and butterflies also have chewing mouthparts, although the adult's mouthparts are highly modified. The diets of chewing insects may be herbivorous or carnivorous, but some diets are highly specific. The specialization of insect mouthparts has been primarily in modifications for piercing and sucking.

Adaptations for the same feeding habits are not uniform, however, because a sucking or piercing feeding habit evolved independently in different insect orders. Moreover, the mouthparts may be adapted for more than one function: chewing and sucking, cutting and sucking, or piercing and sucking. The mouthparts of moths and butterflies are adapted for sucking liquid food, such as nectar, from flowers. A part (the galea) of each of the two greatly modified maxillae forms a long tube through which food is sucked.

When the insect is not feeding, the tube is coiled. The other mouthparts are absent or are vestigial. Piercing mouthparts are characteristic of herbivorous insects, such as aphids and leafhoppers, which feed on plant juices. Some predaceous insects, such as assassin bugs and mosquitoes, which utilize the body fluids of other animals as food, also have piercing mouthparts. In all these insects the mouthparts are elongated and are organized in various ways to form a beak.Most insects possess a pair of salivary, or labial, glands, that lie below the midgut and have a common duct opening into the buccal cavity.

The insect nervous system is basically like that of other arthropods. The brain is composed of a protocerebrum with eyes, deutocerebrum with antennae, and a tritocerebrum. The ventral nerve cord forms a chain of median segmental ganglia. As in other arthropods the ventral segmental ganglia, both thoracic and abdominal, are often fused. The greatest number of free ganglia is three in the thorax and eight in the abdomen (apterygotes). The subesophageal ganglion is always composed of three pairs of fused ganglia, which control the mouthparts, the salivary glands, and some of the cervical muscles.

The typical female reproductive system consists of two ovaries, one on each side, and two lateral oviducts. Each ovary consists of a group of tubules, tile ovarioles. The paired oviducts usually unite to form a common oviduct, which leads into a vagina. The vagina in turn opens onto the ventral surface behind the seventh, eighth, or ninth segment. Diverticula of the common oviduct or vagina include a spermatheca (seminal receptacle) and paired accessory glands. Some Lepidoptera have a separate copulatory gonopore located behind the vaginal opening. The copulatory gonopore leads to a bursa copulatrix that receives the spermatophores.

Sperm reach the spermatheca by a special duct from the bursa to the vagina. The male reproductive system includes a pair of testes, a pair of lateral ducts, and a median duct opening through a ventral penis, called the aedeagus, associated with the eighth segment. Each testis consists of a group of sperm tubes containing spermatozoa in various stages of development. These tubes empty into a lateral duct, the vas deferens, which joins the duct from the other side to form a common ejaculatory duct. A section of each vas deferens is usually enlarged into a seminal vesicle, where sperm are stored.

Many insects mate only once in their lifetime, and none mate more than a few times. When the eggs reach the oviduct, they are already surrounded by a shell-like membrane (chorion) secreted by ovarian follicle cells. This shell may be up to seven layers thick and is perforated by one or more micropyles. The sperm enter through these minute openings. The evolution of a protective encasement for the development of the egg in a desiccating terrestrial environment is another factor that has contributed to the success of insects.

Protochordates

The Hemichordata, Urochordata, and Cephalochordata, were classified as chordates and were seen to embody clues to the early evolutionary history of the phylum. When the homology of the chordate notochord and a similar structure in hemichordates was thrown into doubt, the hemichordates were removed from the Chordata and isolated in their own phylum.

Despite this discrepancy, the biology of hemichordates continues to be a source of information used in recent models of chordate evolution. For this reason, the hemichordates, cephalochordates, and urochordates are lumped together here under the informal heading of protochordates, or "first chordates."

Phylum hemichordata

Hemichordates are large solitary worms (Enteropneusta) or hydroid-like colonies of tiny zooids (Pterobranchia) inhabiting the sea.

The bodies of hemichordates are divided into three regions-a locomotory preoral lobe, a short collar, and a trunk-reflecting an underlying tricoelomate organization. A complex heart/kidney occurs in the preoral lobe and is supported by a notochord-like structure. The principal nervous concentration occurs dorsally in the collar region and, in the Enteropneusta, is hollow. A pharynx, which bears one or more pairs of gill slits, is found in the trunk.

The occurrence, or rudimentary expression, of three chordate features—notochord, dorsal hollow nerve cord, and gill slits—was formerly the basis for including the hemichordates within the chordates. This is no longer an accepted classification, but the hemichordates nevertheless are close relatives of the chordates and provide many clues to their functional organization and ancestry.

Class enteropneusta

Enteropneusta, or acorn worms, are primarily benthic inhabitants of shallow water, but some occur in the deep sea and in association with hydrothermal vents. Some live under stones and shells, and many common species, including those of Saccoglossus and Balanoglossus, construct mucuslined burrows in mud and sand. Exposed tidal flats are frequently dotted with the coiled, ropelike castings of these animals. Acorn worms are relatively large animals, the majority ranging between 9 and 45 cm in length. Saccoglossus (Dolichoglossus) kowalevskii is typically 15 cm long, and the common Balanoglossus aurantiacus of the southeastern coast of the United States reaches approximately 1 m in length.

The largest described acorn worm, Balanoglossus gigas, which ranges from Brazil to Cape Hatteras, North Carolina, exceeds 2.5 m in length and constructs 3 m-long burrows. The cylindrical and rather flaccid body is composed of an anterior proboscis, a middle collar, and a long posterior trunk. These regions correspond to the typical deuterostome body divisions: protosome, mesosome, and metasome. The proboscis is usually short and more or less conical and is connected to the collar by a slender but sturdy stalk. The collar is a short cylinder that anteriorly overlaps the proboscis stalk, embraces the posterior end of the proboscis, and contains the mouth on the ventral side.

In some species, the combined proboscis and collar resemble an acorn and its cap, and the common name-acorn worm-was derived from this similarity. The trunk constitutes the major part of the body and is divisible from anterior to posterior into three regions: the branchiogenital region, the hepatic region, and the intestinal region. The branchiogenital region bears a longitudinal row of gill pores on each side of the dorsal midline. Lateral to the two rows of gill pores, the trunk bears either two low genital ridges (Saccoglossus, Schizocardium) or two large flaplike genital wings (Balanoglossus, Ptychodera), which arch over the gill region and enclose water channel.

The enteropneust epidermis lacks a cuticle but is densely ciliated and richly glandular, and bears numerous short microvilli. The gland cells, which are especially abundant at the tip of the proboscis and on the collar, are responsible for secreting the mucus covering of the body. Some of the gland cells produce bromine-containing compounds that impart a strong medicinal odor to acorn worms and may protect them from bacterial infection or animal predation. A basal lamina and a connective tissue dermis lie beneath the epidermis, but the dermal fibers are not well developed or well organized, perhaps accounting for the fragility of the body. Acorn worms are tricoelomate, like most lophophorates and echinoderms. A single unpaired protocoel occupies the proboscis, a pair of mesocoels is found in the collar, and a pair of metacoels is found in the trunk. The protocoel opens to the exterior on the proboscis stalk via one small opening, the proboscis pore. Each of the mesocoels has a duct that leads to the exterior by way of the first gill pores; the metacoel lacks coelomic ducts.

As in echinoderms and most chordates, there is a trend in hemichordates for specialized diverticula to develop from one coelomic region anti project into another. In enteropneusts, the clearest example of this phenomenon is the extension of two tubular anterior outgrowths of the metacoels, the perihemal coeloms, into the collar region of the body, one on each side of the dorsal blood vessel. The linings of the perihemal coeloms are muscular, and their action may help to circulate blood in the dorsal blood vessel.

The general coelomic lining is a simple noristratified mesothelium, and most of the cells are differentiated as smooth muscle cells, each bearing a single cilium, which may help to circulate the coelomic fluid. The muscle cells are pseudostratified into a poorly developed outer circular musculature and an inner, well-developed, longitudinal musculature. In addition, radial muscle fibers span the coelom in all body regions and may functionally replace the weak circular muscles.

The Musculature is best developed in the proboscis, the locomotory organ, and in the trunk, where two ventrolateral bands form the retractor muscles that shorten the trunk. Generally, acorn worms are sluggish animals that move slowly using retrograde peristaltic contractions of the proboscis. After the proboscis advances and anchors, the trunk and collar are pulled forward and the cycle is repeated.

Acorn worms may be deposit feeders, suspension feeders, or both. Burrowing species are predominantly deposit feeders, consuming sand and mud and digesting out the organic matter, but a few species protrude the proboscis from the burrow opening and collect suspended material from the water. Nonburrowing species are primarily suspension feeders but may also ingest deposited material. Food in the form of detritus and plankton, or sediment in burrowing species, is trapped in mucus on the proboscis and transported posteriorly over its surface toward the mouth.

Two common burrowing enteropneusts, Saccoglossus kowalevskii and Balanoglosus aurantiacus, are both surface deposit feeders that occupy U-shaped burrows but have markedly different feeding styles. Saccoglossus emerges from its burrow and extends its long proboscis radially over the surface from the burrow opening. While extended, the proboscis conveys the surface detritus back to the mouth.

After feeding for a time, the proboscis is retracted and then re-extended in a new direction. After several cycles of proboscis movements, Saccoglossus creates a spoke like series of radiating furrows, a feeding rosette, around the mouth of its burrow. The burrow of Balanoglossus typically has two surface openings and a third crater-like depression, the feeding funnel, which lies above a branch from the head end of the U-shaped burrow. While feeding, Balanoglossus occupies the branch and positions its proboscis immediately below the bottom of the surface crater. As it removes sand from the base of the funnel, surface detritus tumbles into the funnel, falls to the bottom, and is ingested by the worm. Thus, Balanoglossus is able to feed on the organically rich surface detritus without exposing any part of its body above its protective burrow.

The blood-vascular system of hemichordates consists of an anterior heart and two main contractile vessels, one middorsal carrying the colorless blood forward, and one midventral, carrying blood posteriorly. Smaller vessels supply the body wall and major organs. The heart/kidney is located in the posterior of the protocoel. It is supported beneath by the stomochord and rigid proboscis skeleton and pressurized above by a small contractile coelom, the pericardium.

The physiology of excretion in hemichordates has not been studied. The pharyngeal gill slits are assumed to be gas exchange structures. The number of slits can range from a few to as many as 200 or more pairs. Each slit opens through the side of the pharyngeal wall as a narrow U-shaped cleft. The pharyngeal wall between the clefts (the gill arches) and that part of the wall projecting downward between the arms of the U (the tongue bar) are supported by skeletal thickenings of the dermis of the pharyngeal epithelium.

Acorn worms are exceptionally fragile animals and invariably tear or break posterior to the branchiogenital region when handled. It is likely that they also suffer damage under natural conditions and then heal the wounds and regenerate missing parts.

Asexual reproduction by fragmentation has been reported for several species, including members of the genera Glossobalanus and Balanoglossus. Acorn worms are dioecious animals that release eggs and sperm into the surrounding seawater where fertilization occurs. The numerous follicular gonads are situated in the connective tissue space of the genital ridges (Harrimaniidae and Spengelidae) or genital wings (Ptychoderidae). Each follicle has its own surface pore. Male gonads are often orange and female gonads, lavender or grey. The zygotes undergo holoblastic, equal, radial cleavage.

Class pterobranchia

Pterobranchs are colonial, tube-dwelling, epibenthic hemichordates that superficially resemble bryozoans and hydroids. As in these other two groups, pterobranch zooids are small, not exceeding 5 mm in body length, and many species are less than 1 mm long. But despite their small body size, pterobranch organ systems are similar to those of their larger relatives, the enteropneusts. Because of these similarities, only the distinctive features of pterobranchs will be discussed in the following section.

Pterobranchs are uncommon marine animals that rarely occur in shallow water where they grow on the surfaces of shells and rocks. More often they are dredged from deep water and are best known from the seas around Antarctica. There is growing evidence, however, that pterobranchs are widespread, but easily overlooked, in shallow tropical and temperate seas, including the coasts of Florida and Bermuda.

There are 21 species of pterobranchs in three genera: Cephalodiscus, Atubaria, and Rhabdopleura. Atubaria, however, closely resembles a Cephalodiscus that has abandoned its tube, which it often does, and is known only from preserved collections dredged from a depth of 200 to 300 m off the coast of Japan. Although pterobranchs are rarely seen by most biologists, they have nevertheless figured prominently in theories of deuterostome evolution. Most models of the evolution of echinoderms and chordates place a pterobranch-like ancestor at the base of the evolutionary tree, and within the hemichordates, the pterobranchs are primitive to the enteropneusts in most of their shared structural traits.

The tripartite body consists of an anterior oral shield, a short collar, and a saccate trunk. The broad, flattened oral shield is a glandular, ciliated disc on which the zooids glide over the inner surface of their tubes. The short collar region arches dorsally over the oral shield and bears the arms and their pinnately arranged tentacles. Pterobranch colonies inhabit a branching tubular network, called a coenecium, that is secreted from glands in the oral shields of the zooids.

In Rhabdopleura, the coenecium consists of prostrate tubes that give rise to a series of unbranched vertical tubes, each of which has a simple circular opening and is occupied by a single zooid. The stalks of all zooids in the colony join a common threadlike stolon that runs through the prostrate tubes. Most colonies of Rhabdopleura are between 1 and 25 cm in diameter and 5 to 10 mm in height.

The coenecium of some species of Cephalodiscus resembles that of Rhabdopleura but is typically larger. A Cephalodiscus coenecium is occupied by many unjoined zooids, each of which produces buds from an adhesive disc at the base of the stalk to form a small colony.

The pterobranch body wall is similar to that of enteropneusts, but in pterobranchs, the cells are always monociliated, a primitive feature. The musculature consists of longitudinal and radial fibers only; circular body wall muscles are absent. The radial muscles are common in the oral shield, collar, and arms. Smaller blood vessels are known to occur, but details of the blood circuit and direction of flow of the colorless blood are unknown. All blood vessels and sinuses are simple, unlined channels in the connective tissue between epithelia. Gas exchange most likely occurs across the arms, tentacles, and general body surface, all of which are ciliated and generate water flow over the surface of the body. The nervous system is entirely within the epidermis, and a hollow, submerged collar cord is absent, although a concentration of neurons occurs dorsally in the collar region of the body.

Pterobranchs filter feed using their hollow ciliated arms and tentacles. Lateral cilia on the tentacles create a flow of water and particles over the frontal surface of the tentacles and arms. Small particles are intercepted by these frontal surfaces or are batted onto them by the flicking tentacles. Once captured, cilia transport the particles along the length of the arms to a ciliated groove that passes under the undulating oral lamellae before joining the mouth.

Pterobranch zooids arise asexually, by budding from the stolon or adhesive disc during colony growth. Moreover, an individual adhesive disc of Cephalodiscus may be capable of fragmentation or fission to form new colonies. Pterobranch colonies are believed to be dioecious, and the male and female zooids, which have one (Rhabdopleura) or two (Cephalodiscus) gonads, often show sexual dimorphism. In Cephalodiscus sibogae, for example, male zooids have a rudimentary lophophore, and much of the trunk is occupied by the large testes.

Chapter 2

Cellular and Extracellular Environment of Organisms

The cell

Cells are living units of an organism both structurally and functionally. Each has its own internal complement of functional parts and its own independent physiology. Generally similar internal structures play like roles in the living processes of all cells. In all but the lowest of multicellular animals, cells of similar specialized types combine to form similar tissues, and these tissues combine to form similar organs.

No organ system, organ, tissue, or cell operates alone or independently of the rest of the animal in all ways, but many cellular functions are carried out independently by each cell. A typical animal cell, however, contains a number of recognizable parts or organelles. Typically there is a prominent nucleus that contains nucleoplasm and is enclosed by a nuclear membrane. The remainder of the cell consists of cytoplasm, and the whole cell is enclosed by a plasma membrane. The nucleoplasm and cytoplasm, together with their membranes and inclusions, are collectively spoken of as protoplasm. In other words, protoplasm is the general name for all living matter. Protoplasm is largely composed of water about 70 to 80 per cent, but contains in addition electrolytes, proteins, carbohydrates, lipids, and various other organic materials.

Cell organelles

Plasma membrane

The cell membranes are most probably, semipermeable, and selected and controlled movements of various substances occur through them. The structure of all cell membranes is, apparently, quite similar but of sufficient complexity to elude, thus far, positive understanding. The structure for the plasma membrane originally suggested by Danielli was later applied by others to all cell membranes and referred to as the unit membrane. This concept pictured all cell membranes as having, fundamentally, a double layer of lipid molecules overlaid on both sides by a layer of protein molecules.

The outermost layer of the plasma membrane, at least, is now known to contain a considerable amount of carbohydrate. Except for this, however, the Danielli membrane composition is still generally accepted.There is abundant evidence to suggest that typical cell membranes have tiny pores penetrating through them. Moreover, these pores can apparently vary in size from time to time. Such variation in pore size results in changes in the permeability, of the membrane to various substances.

Attempts to explain the changing porosity of membranes have led to a somewhat altered version of the Danielli membrane structure. According to this modern concept, the lipid portion of the membrane consists of a number of "pillars" with pores between them. Variations in the shape of the pillars results in changes in the pore size. The plasma membrane enclosing the entire cell is almost certainly of the type just described.

Endoplasmic Reticulum (ER)

Cytoplasm of the typical cell is filled with a series of interconnected membranes collectively referred to as the endoplasmic reticulum (ER). These membranes form tubules, spaces, and vesicles that are filled with an endoplasmic matrix. The endoplasmic matrix is a fluid different in composition from the general cytoplasmic fluids that surround these spaces.

Some portions of the endoplasmic reticulum are marked by the presence of small granular particles called ribosomes. These particles are apparently attached to the endoplasmic reticulum, giving it a roughened appearance. This portion of the endoplasmic reticulum is accordingly called the granular, or rough, ER. Other portions lack ribosomes and are differentiated by the name smooth ER.

Golgi complex

Portions of the smooth ER are expanded to form larger vesicles referred to as the Golgi complex. The Golgi complex is most prominent in secretory cells of various kinds. It apparently functions in the temporary storage of materials that are to be secreted. The membranes forming the endoplasmic reticulum and Golgi complex are continuous with the nuclear membrane and, perhaps, with the plasma membrane as well. It now seems certain that the rough portions of the endoplasmic reticulum function in the synthesis of proteins from amino acids.

Ribosomes

Ribosomes, in turn, are constructed in the nucleus of the cell under the control of the genetic material of the chromosomes. The smooth portions of the endoplasmic reticulum probably function in the synthesis of non-protein materials, especially lipids and various secretion products of cells. The secretion products may also include proteins formed on the rough ER. In either case, they are then transferred to the Golgi complex for storage, concentrated to form secretion granules, and finally passed from the cell.

Mitochondria

Variable numbers of mitochondria occur in the cytoplasm of all animal cells. They are present in greater numbers in cells with larger energy requirements and, from this as well as from other evidence, the mitochondria have been identified as the "powerhouses" of the cell. Sugar can be broken down in the cytoplasm to yield some energy and smaller hydrocarbon molecules. This process is known as anaerobic glycolysis and involves a series of biochemical reactions known collectively as the Embden-Meyerhof pathway. In the mitochondria these smaller hydrocarbons are oxidized to form carbon dioxide and water and to yield much larger amounts of energy. These reactions within the mitochondria involve a cyclic biochemical process known as the Krebs cycle and a series of cytochrome enzymes.

All the many steps of these reactions require specific organic catalysts (enzymes) in order to proceed at appropriate rates. It is thought that these enzymes are arranged sequentially on the innermost of the two mitochondrial membranes. This inner membrane is thrown into folds, or cristae, which greatly increase its surface area.

Lysosomes

Many cells contain accumulations of enzymes that are enclosed within membranes to form structures called lysosomes. In one-celled animals the lysosomes function by discharging their contained enzymes into food vacuoles. There the enzymes work by catalyzing the molecular breakdown of food materials so that the products can be absorbed into the cytoplasm of the cell. Thus, they act as important parts of the digestive "system" of these one-celled animals. In vertebrate animals some of the lysosomes function in the same general way since their contained enzymes are discharged into the digestive tract. Lysosomes probably also act in extensive tissue remodeling processes, such as the metamorphosis of amphibians, the resorption of bone by osteoclasts, and the atrophy of unused muscles. The lysosomes are undoubtedly derived from the Golgi complex, in which the newly synthesized enzymes are first stored.

Centrioles

Generally, a typical cell contain two small structures called centrioles—that play important roles in cell division. Each is a cylinder composed of nine parallel fibrils. The two centrioles generally lie at right angles to each other in resting cells. At the beginning of prophase of cell division one centriole migrates to each pole of the dividing cell and there acts as a center for the organization of the spindle fibers. Following the completion of cell division, one centriole remains in each of the two daughter cells.

Cellular outgrowths

Some cells have extensions called cilia or flagella, and these contain nine parallel fibrils, as do centrioles. They are called the outer fibers of the cilium or flagellum and are enclosed by an extension of the plasma membrane of the cell. Each outer fiber, like those of the centrioles appears to be doubled when seen in cross section. In addition, cilia and flagella have a doubled central fiber and sometimes additional fibers as well.

At the base of each cilium or flagellum is a structure called the basal granule. It lies just beneath the cell surface and has a structure much like that of a centriole, i.e., the nine fibrils are present but there is no central fiber. Moreover, the nine outer fibrils appear to be tripled rather than doubled. It is assumed that the basal granule functions in some way to control the activity of cilia and flagella and that the actual contractile elements lie within the cilia and flagella themselves. Cilia that function in cell motility or in propelling environmental media past fixed cells are found in respiratory passages, sinus cavities, oviducts, and elsewhere in vertebrate animals.

Vacuoles and crystals

Cells often contain other identifiable structures such as vacuoles and crystals. Vacuoles are usually enclosed by membranes derived from the plasma membrane and contain materials engulfed from the surrounding medium or produced by the Golgi complex of the cell. Specialized cell types contain still other organelles; for example, the contractile fibers and microtubules present in muscle cells.

Nucleoplasm

Nucleoplasm consists of the genetic materials of the chromosomes as well as proteins and small amounts of other substances. Morphologically nuclear inclusions are more variable than those of the cytoplasm not only from cell to cell, but within the same cell at different times. Chromosomes as such are seen clearly only during active cell division. The most frequently identifiable structure present, other than the chromosomes, is the nucleolus. This is a small clump of ribonuclear protein lacking an enclosing membrane and attached to two or more of the chromosomes.

Types of animal cells

The cells in animal body are differentiated in to variuos types according to their function.

Epithelial cells

The cells of epithelial tissues are arranged in single layers (simple epithelium) or in multiple layers (stratified epithelium). Consequently, they occur in various shapes ranging from flattened plates (squamous cells) to cuboid and columnar forms. In simple epithelium the cells tend to be polarized; that-is, the superficial end of the cell is different in appearance from the deep end. The deep end rests on a basement membrane or lamina, which is composed of extracellular materials and separates the epithelium from deeper tissues.

The superficial, exposed end of epithelial cells may bear cilia or they may have a so-called brush border. The striated appearance of a brush border results from the presence of numerous cell processes known as microvilli. These differ from cilia in that they do not contain the characteristic fibrillar structures. In still other cases the superficial ends of epithelial cells are covered with a tough extracellular material known as keratin. Epithelial cells, which are glandular in nature, produce secreted materials that are discharged onto the surface of the tissue. Such secreted materials include protective and lubricating mucus as well as digestive enzymes.

Glandular cells of the lining of the digestive tract produce both kinds of secretions. Sometimes glandular cells occur singly and sometimes in groups, forming organized glands. Glandular epithelial cells contain many mitoehondria and a well-developed Golgi complex; this would be expected from their high level of activity.

Connective tissue cells

Connective tissues include bone, cartilage, tendons, ligaments, and the finer tissues that bind together other kinds of tissues. Blood is commonly classed as a connective tissue, even though it does not play the same kinds of roles, because of similarities between the embryonic origins of blood cells and other types of connective tissue.

More typical connective tissues consist of cells plus greater or lesser amounts of extracellular materials secreted by those cells. Extracellular materials consist of fibrous or elastic strands, deposits of a homogenous matrix, or both. The matrix may be soft or semifluid as in hyaline cartilage, or it may be hard and mineralized as in bone.

The cells in cartilage and bone lie comparatively far apart in spaces or lacunae with a matrix filling between the cells. In other kinds of connective tissue such as tendons, ligaments, and fascia, there are more fibrous or elastic strands, less homogenous matrix, and the cells lie scattered throughout the tissue. Generally, there are three types of bone cells: (1) osteoblasts, which Jay down the mineralized matrix of hydroxyapatite ($Ca_{10}(PO_4)_6(OH)_2$); (2) osteocytes, which lie "trapped" in the lacunae of the completed bone; and (3) osteoclasts, which lie on the surfaces of bone and cause resorption of the mineralized matrix.

Osteoblasts and osteoclasts working together make possible the continual remodeling of bone structure that accompanies growth and development and that adapts completed bones to new stresses. Tendons, ligaments, and fascia contain cells known as fibroblasts. These are spindle-shaped (fusiform), flattened, or stellate cells which form the fibrous or elastic strands of such tissues. Also present are macrophages, which are phagocytic cells; that is, they are capable of engulfing in ameboid fashion various kinds of foreign materials or cell detritus. Some macrophages are fixed in position and some are wandering, cells capable of moving about through the tissues like amebae. Adipose cells, or fat cells, are also classed as connective tissues. These function in the storage of fats and consist of little more than large vacuoles enclosed by a plasma membrane. A flattened nucleus lies at the edge of the cell.

Musclular cells

Muscle tissue makes up the greatest amount of the soft tissues of most vertebrates. The physiology of contraction has received a great deal of attention from investigators.

With the advent of the electron microscope the fine structure of skeletal muscle cells became known, and it was possible to draw tentative conclusions regarding the mechanics of contraction. Less is known of the physiology of smooth muscle, but there is evidence to suggest that both types are fundamentally similar. Indeed, it may be that all protoplasmic movement is based upon similar biochemical processes but is variously modified by the different mechanical orientations of cell parts.

The cells of skeletal muscle are comparatively large; they may range up to several centimeters in length and from 10 to 100 microns in diameter. Each cell is called a muscle fiber, and groups of cells are bound together by connective tissues to form fasciculi. Inside each skeletal muscle cell are several irregularly spaced nuclei that are otherwise not very different in appearance from typical cell nuclei. Skeletal muscle cells are strikingly different, however, in that they contain large numbers of longitudinal myofibrils which, when seen through the optical microscope, have prominent cross-striations. The electron microscope reveals that the myofibrils are, in reality, composed of filaments of two general diameters, thick and thin. These interdigitate with each other in a repeating pattern, and it is this orderly arrangement of filaments that gives the striated appearance to the myofibrils. Each unit of the pattern is called a sarcomere and terminates at each end at a Z disc. The thinnest of the filaments are attached to the Z discs, whereas the thicker filaments are suspended, apparently unattached, between them.

Chemical extraction has shown that the thick filaments are composed of a protein called myosin and that the thin filaments contain a different protein called actin. Chemical reactions between these two proteins during contraction cause the two kinds of filaments to interdigitate further with each other thus inducing a shortening of the sarcomere. Since all of the sarcomeres of the cell shorten simultaneously, the entire cell shortens. The energy-storing molecule adenosine triphosphate, present in large amounts in skeletal muscle, is associated with the actin filaments, and there are numerous mitochondria within the cell. The endoplasmic reticulum of skeletal muscle cells is highly specialized and has been given the name sarcoplasmic reticulum. It consists of longitudinally oriented microtubules that lie between and parallel to the myofibrils of the cell.

In the region of the interdigitation of filaments these tubules are associated with enlarged vesicles called triads. Nervous impulses are electrical in nature and are transferred to the membrane of skeletal muscles at their points of innervation. It is believed that the sarcoplasmic reticulum carries these electrical disturbances throughout the muscle mass and that this causes a release of calcium ions at the triads. The calcium ions, in turn, trigger the internal chemical events associated with actual contraction of the cells.

Nerve cells

Nerve cells are designed to transmit nervous impulses from place to place, but others apparently serve as structural elements to hold portions of the system together, or there may function in still different roles. Indeed, we are at present a long way from understanding the functions of a number of cell types within the central nervous system. There is really no "typical" neuron, but we may base our discussion, on a generalized conducting neuron of the type located in the longer central nervous tracts or outside the central nervous system. Such a neuron is composed of a cell body, a few short dendrites, and a single long axon. Ordinarily a dendrite picks up an impulse from a sensory receptor or from another neuron; the impulse is then transmitted through the cell body and along the axon to an effector organ or to another neuron.

The point of contact between one neuron and another is called a synapse, and much of the specificity and versatility of the nervous system is due to the properties of its many synapses. Compared with other cell types the length of a single neuron is sometimes very great. For example, a motor neuron passing from the spinal cord to muscles in the foot of a man may be more than three feet in length. If the cell body were enlarged to the size of a tennis ball, the axon would then be a little less than a half-inch in diameter and about a mile in length. The cell body contains a large nucleus and variable amounts of chromatin material scattered in the form of granules.

Other typical organelles, such as the Golgi complex, mitochondria, and centrioles, are also present. The cell body and the dendrites and axon also contain a network of fine neurofibrils. Neuron processes lying outside the central nervous system are often covered with complex layers of a fatty substance called myelin and a membrane called the neurilemma or sheath of Schwann. These are wrapped about the axon in a spiral fashion because of the way in which the Schwann cells wrap extensions of their plasma membranes about the neuron process.

Sex cells

The sex cells, are highly specialized cells containing only half as much nuclear genetic material as other cells. The female sex cells, or ova, are relatively large, spherical cells containing variable amounts of yolk material and often surrounded by numbers of protective follicular cells.

The male sex cells, or spermatozoa, are more complex. In these, the genetic materials (nucleus) are encased in a head, which often has a protective cap lying over it. Attached to the head is a tail consisting of a neck, middle piece, principal piece, and end piece. The middle piece has a sheath wrapped about it that also encloses numerous mitochondria. The tail contains the typical get of fibrils found in cilia or flagella, and indeed this is a highly specialized flagellum. The shape of the head varies greatly in different vertebrate species.

Cell-created environments

Majority of cells in multicellular organisms have fixed positions relative to one another. Circulating red and white blood cells and wandering macrophages are examples of cells that are free from these constraints, but most others are cemented into a matrix that they themselves have created. The amount of matrix between adjacent cells varies with different tissue and organ types. For example, the matrix width between cells in epithelial tissue is minimal. There are even points of intercellular membrane fusion, eliminating the extracellular environment completely. In cartilage, cells are widely spaced in a matrix of much greater volume than that occupied by the cells.

Collagen

The most abundant substance in matrix material is collagen. More than 70% of the dry weight of tendon, skin, and cornea is collagen. This protein is actually a family of gene products rather than a single type of protein molecule, so it is appropriate to use the plural, collagens, when discussing these systems. The native collagen molecule consists of three polypeptide chains Wound to form a triple helix. At least five types of collagen molecule are recognized, each made up of a different combination of a chains. Type I for example, consists of two al(l) chains and one a2(l) chain. This type of collagen appears in some of the harder tissues, such as bone and tendon. A second type of a1 chain, designated II, is found in triplicate in the collagens of cartilages. Electron microscopic studies on the collagens reveal them to be striated fibers with repeat distances of 640 A. This morphology is best explained by a staggered arrangement of end-to-end tropocollagen molecules, each tropocollagen being the elemental unit of the intact collagen fiber. The amino acid composition of the collagens is quite compared to other proteins. The a chain, for example, contains about 1050 amino acids, with glycine contributing about one third of the total residues.

The primary structure is thus $(\text{—Gly—X—Y})_n$ where X and Y may be filled by a variety of amino acids. Most commonly, however, the X-position is taken by proline and the Y-position by hydroxyproline. Collagens are synthesized by cells in *procollagens.* The mechanism of cell synthesis and secretion follows the same pathway as that utilized for any other secretory protein, namely, synthesis on ribosomes of the rough endoplasmic reticulum followed by transport through the Golgi complex and exocytosis from the cell.

The procollagen molecule, when secreted, is already a hydroxylated and glycosylated triple helix molecule. Proline and lysine are both hydroxylated, with about 10% of the proline and 0.5 to 4% of the lysine in hydroxylated form. The hydroxylation reactions proceed by specific hydroxylase enzymes operating on the membrane-bound ribosomes.

Carbohydrates are covalently attached to the collagens through the side chains of hydroxylysine. Glycosylation takes place in the cell at the level of the nascent peptides by UDP-galactosyl transferase and UDP-galactosyl transferase enzymes. Thus there are four intracellular enzymes that effect a posttranslational modification of the collagens, two concerned with hydroxylations and two with glycosylations.

The assembly of the triple-helical configuration also takes place before secretion. It is now thought that the pairing of the a chains is promoted by extensions called *registration peptides,* which are eventually excised from the procollagen before the collagen fiber is formed. Concomitant with triple helix formation, disulfide bonds are formed between the chains, shown here toward the C-terminal ends of the chains. The helix is thereby stabilized by covalent linkages and in this form procollagen is secreted from the cell. The amino-terminal registration peptides are cleaved off first, leaving a processed procollagen that still contains the carboxy-terminal stretches cross-linked by disulfide bonds. These stretches such sequently excised to generate the tropocollagen molecule. The Collagen fibril is then formed by a self-assembly process in which tropocollagen molecules aggregate. The fibril is finally strengthened by covalent linkages employing oxidized forms of lysine and hydroxylysine that condense.

Two adjacent tropocollagen molecules may be cross-linked by an aldol condensation between their oxidized lysines. Another molecule may be tied to the complex by forming a Schiff base between the free aldchyde and an amino group of lysine. Still another tropocollagen can be bound by a reaction between an imidazole group of histidine and the unsaturated carbon—carbon bond. Monosaccharides, such as glucose and galactose, that are destined to be incorporated into polymers are generally activated and carried by UDP. One or more of these reactions may take place to increase the tensile strength of the collagen fiber.

Although the number and variety of cross-linkages varies with the type of collagen, the linkages are usually grouped toward the end of the tropocollagen molecules. These, then, are the types of molecule that make up a good deal of the material in the extracellular matrix in tissues and organs. The full impact of matrix collagen on the cell is not yet realized, but (in extensive amount of research is currently underway to ascertain its effects upon cell properties. Collagen enhances Such cell properties as attachment, growth, and differentiation. It is likely that these cell activities are affected similarly *in vivo*. Furthermore, different types of collagen influence cells differently.

Epidermal cells attach best to type IV collagen, whereas fibroblasts adhere equally to all collagen types. The attachment of fibroblasts to collagen, although not influenced by collagen type, is enhanced by a protein that has been given the name *fibronectin*. This protein apparently acts as a mediator between the cell and the matrix.

The attachment of epidermal cells, on the other hand, is not enhanced by fibronectin. It therefore appears that matrix-cell adherence is governed by the type of matrix molecules and by mediators. These observations lend support to the notion that during development cell localization and orientation may depend on specific interactions with matrix molecules.

Proteoglycans

Proteoglycans consist of proteins and glycosaminoglycans (or mucopolysaccharides) linked both covalently and noncovalently to form large amorphous aggregates. The proteoglycan aggregate is presently viewed as having a structure. The backbone of the aggregate is an extended hyaluronic acid polymer, highly acidic because of its content of carboxyl groups on every other sugar unit. Around this backbone proteoglycan subunits are clustered, held in place by noncovalent associations. The proteoglycan subunit appears to consist of three regions, a globular end that binds noncovalently to the hyaluronic acid, an intermediate length of molecule to which keratan sulphate is covalently attached by way of glutamic acid residues, and an extended region of variable length where both keratan sulphate and chondroitan sulphate are covalently bound.

Another protein, called a link protein, is present in these aggregates. Its function appears to be to stabilize the noncovalent association between the proteoglycan subunits and the hyaluronic acid backbone. In articular cartilage, where the cells are widely separated by large amounts of extracellular matrix, the proteoglycan aggregate makes a very significant contribution to the extracellular material. In this cartilage, the matrix consists of approximately 70% collagen, 18% proteoglycan, and 10% additional noncollagenous protein. Proteoglycans interact with collagen molecules to varying degrees, depending on their composition and the composition of chains making up the collagens. An area of researeh that is now receiving concentrated effort is the relation between these components during development.

Basement membranes

Basement membranes are rather a specialized form of connective tissue made up of extracellular matrices rich in type IV collagen, not in any way related chemically or structurally to the plasma membrane. Basement membranes are formed by the cells with which they are located. For example, the glomerular basement membrane is made by the epithelial cell, and so is the lens capsule and basement membranes in the vicinity of respiratory, glandular and intestinal epithelium. In like manner, Descemet's membrane is produced by enclothelial cells and Reichert's membrane of the embryonic yolk sac by endodermal cells. Basement membranes have served well as model systems for the study of the extracellular environment. They can be isolated, with those from the glomerulus and the lens more frequently isolated and studied than the others. Their amino acid composition and carbohydrate content have been explored, and certain physical properties, Such as Solubility, have been examined. From these studies it can be concluded that basement membranes contain only peptide and carbohydrate.

Basement membranes from different sources are qualitatively similar but quantitatively distinct. There is characteristically a high content of proline and hydroxyproline, glycine, and hydroxylysine. The carbohydrate content of these extracellular matrices is high, ranging between 11 and 14%. Hexoses are predominant, with the presence of glucose setting basement membranes apart from most other tissue and plasma proteins. One of the proteins of basement membranes is clearly a type of collagen that contains glucose and galactose almost exclusively which together form glucosylgalactose disaccharides distributed on the a chain.

In addition to the collagen a high molecular weight glycoprotein is present in basement membranes. Basement membranes probably have at least two functions. One function is to provide an elastic form of support to epithelial and endothelial cell linings. It is a medium on which these cells are anchored, thereby enabling them to congregate into issue form. A second function appears to be that of filtration. Molecules that move across the epithelial surface into the blood and ultimately across the capillary walls must make their way through basement membrane material.

Extracellular environment of organisms

The sea constituted the prime seed bed for living things. Cellular diversification would not have been necessary at least initially, since the sea provides an enormous volume of habitat with a relatively fixed composition. As cellular life began to probe the habitat potentials of new environments the ability to adapt depended on the genetic machinery and ability to change. Change resulting in an advantage often was accompanied by the loss of certain abilities. Thus cells began to specialize ace to the properties of their environments, with some tolerating lower salinity and others adjusting to different climates of temperature pressure, oxygen, and pH. As unicellular life realized an advantage in multicellular existence further specialization took place. Cells began to interact not only wit environment, but with one another. Often this was accomplished creation of connecting materials, making feasible the grouping specialized cells that could interact indirectly by way of extracellular matrices.

Thus not only are cells governed in part by the extracellular material in which they are immersed, but they have become dependent another. They, in turn, either individually or collectively, frequent generate their own environment by synthesizing and secreting matrix material. Procaryotic cells are more versatile than eucaryotic cells in to new environments. They appear to be more versatile biochemically and they have simpler genetic systems that are more readily mobilized take advantage of mutations and the production of adaptive enzymes. They have remained largely unicellular and independent of other cells Furthermore, they have penetrated and established themselves in every conceivable environ-mental niche. Eucaryotic cells have developed highly sophisticated and specialized intracellular hardware, which has made it more difficult for them to adapt because of the complexity of their design. In addition, their use of specialized internal structures has made intercellular dependence of advantage. This type of specialization has promoted the development of tissues and organ systems.

These features have given multicellular organisms an ability to survive in extremely inhospitable environments only as long as their individual cells are sheltered in a microenvironment compatible with life. The necessity for protective coverings, circulatory systems, and mechanisms to maintain an osmotic balance among cells. Every environment has physical or chemical factors that make it compared to any other environment. For an organism or a cell to adapt to a new environment, it must be able to handle a new set of factors, some of which may be lethal. The options available to the cell are few. One option is to keep the new factor out. This means modifying walls or membranes or transport mechanisms. Another option is to modify the factor so that it is no longer a metabolic or structural threat. This inducing enzymes to act on the factor. A third option is to become dependent on the factor. If the factor continues to be lethal for organisms, it will give the dependent organism a competitive advantage in the environment. An environment that is perfectly "normal" for one organism may be extreme to the point of lethality for another. Therefore extremes as we are viewing them are not absolutes, but relative differences in one or more environmental factors that pose hardships to the structure or activities of a given organism. Several factors may contribute toward a condition of environmental extreme. Some of them are described below.

Water environment

All living things have high water contents and depend on the availability of water to carry out the numerous reactions of metabolism. If water is not readily available, metabolism is depressed and so are all the related activities of the organism, including growth and reproduction. One way of viewing water availability is by the concept of water activity, a_w. Two factors commonly influence a_w. One is the degree to which water is adsorbed to surfaces, hence unavailable to the cell: and the second is the extent to which solutes are hydrated in aqueous systems. In either case, if water is not readily available to the organism, its activities will be depressed. Two types of environment in the biosphere contain low water activities. One is the dry environment, where the small amounts of water present are tightly bound to soil particles or other surfaces. The other is the saline environment, where hydrated salts cut down on the availability of water to the organism. Although most organisms grow best at high water activities ($a_w > 0.95$), many are *osmototolerant* and some even *osmophilic.* Those tolerating saline environments, where water activity is decreased, and those loving it, do much better than normal organisms in these environments, as is evident from their growth rates.

Water is an absolute requirement for the cell, both internally and as an extracellular environment. When cells are immersed in pure water or water relatively free of salts, they immediately encounter an osmotic pressure problem. Most cells have an internal ionic environment that represents variations of the theme seen in seawater.

An organism with an internal environment roughly reflecting this composition must be able to prevent an influx water when in a freshwater environment. Without such a mechanism, cells will lyse. Two major cell modifications allow unicellular organisms to survive in low salt, aqueous environments. One is the formation of a cell wall that can resist high osmotic pressure built up within the cell. Bacteria and cyanobacteria possess a wall that can withstand internal pressures of 20 to 30 atm, giving them the option of living quite comfortably in freshwater lakes and rivers, and even in a carboy of distilled water in the laboratory. Other freshwater organisms, such as eucaryotic algae and freshwater fungi, also possess walls permitting life to go on in spite of potential cell-splitting pressures.

A second mechanism to combat low salt environments is employed primarily by protozoans. These organisms have water pumps, internal water-concentrating organelles, which when they are loaded, dump their contents back into the extracellular environment. Protozoans thus keep from cellular lysis by constantly returning the water to the outside.One solution to the osmotic pressure problem is to select or create an environment that contains salts at a concentration that approaches that of the cell interior.

There are two such environments in the biosphere. One is the sea, the largest salt-containing environment available to cells. It is a natural environment. In some sense the original cell environment. It is not generated by cells, and its composition does not have to be monitored or maintained by cells. The second salt-containing environment is blood or body fluids. This is a contrived-extracellular environment that is constantly monitored and maintained by cells to keep it at a composition that is agreeable with life. It is thus a created environment and one on which certain cells depend completely.

The relative proportions of these salts are reflected in most marine invertebrates. Echinoderms, which have no kidneys, have internal fluid-environments most like the seawater in which they reside. Mollusks and crustaceans, possessors of kidneys maintain salt compositions in their body fluids that differ from the sea, being relatively richer in potassium and poorer in magnesium.

All other higher forms of life have body fluids that are lower in salt concentration than the sea. Vertebrates range from 0.68 to 0.9% salt and fresh water and terrestrial invertebrates even lower. A unicellular or small multicellular organism with an internal salt concentration lower than the sea must have cellular mechanisms to prevent water loss and must perform work to extract water from the sea. This implies a specialization of membranes quite different from that needed by protozoans in fresh water, where the problem is to get rid of excess water. Multicellular sea-dwelling organisms with internal environment lower in salt than the sea must depend on the specialized cellular functions of kidneys to enforce these homeostatic levels and to maintain relative compositions of salts that are most favourable for these complex organisms. In addition, the internal salt-containing environment must be protected from the external sea environment because they differ not only in concentration but in composition.

Salinity

As the salt concentration in the environment increases, less water is available for the organism and more work is required to sequester water molecules from the environment. But increased salinity may have other effects on the cell. Certain ions may fie toxic and thus limit the forms that will tolerate them. Some organisms dependent on high concentrations of the ions likely to be found in saline environments. This is true for the organism Halobacterium, which requires relatively high concentrations of sodium ion (Na^+) for growth. It is an example of an extreme *halophile.*

Organisms that live in water live under a pressure exerted by the depth of water in which they are found. For each 10 m of water depth there is an approximate hydrostatic pressure of I atm. Since the hydrostatic pressure in a large body of water is a continuum from top to bottom, it is difficult to say at which depth the pressure becomes extreme.

Some organisms are *barotolerant;* that is, they tolerate high pressures and will grow and reproduce in such environments. These organisms have generally adapted to this environment by possessing enzymes that are not pressure inactivated and by having membranes that are not severely affected by high pressures.It is not now clear whether any organisms are truly *barophilic.* Most barotolerant organisms do better at lower pressures. Thus, they tolerate high pressures, but prefer lower pressures.

pH

Enzymatic activities and structural roles of proteins are highly pH dependent. Even a slight change in the pH, or hydrogen ion concentration, in the environment may adversely affect the metabolism of an organism. In the biosphere acidic environments are more common than alkaline habitats. The ocean is slightly alkaline, pH 8, whereas rivers and takes may have pH values ranging from near neutrality to pH 5 or 6. These are comfortable pH ranges for most organisms.A number of environments are not as comfortable. In bogs and swamps and in volcanic and other geothermal habitats the environment can be highly acidic.

Extreme example of acid environments are solfataras, the sulphur-rich geothermal areas found in many parts of the world. These are basically sulphuric acid environments with pH values that are commonly less than 2. In Yellowstone National Park a solfatara soil with a pH of 0.05 was found to have a living algal organism. Different organisms have different lower pH limits. Fungi tolerate extremely low pH values, as do certain bacterial and algal species. Vascular plants may live at a pH as low as 3 and fish in waters as low as about 4. Because of the relationship between pH and protein structure, it is obvious that aciduric organisms must accomplish one of two things for survival.

Temperature

Organisms, in particular certain microorganisms, are capable of living at enormous temperature extremes. Some live, metabolizing and reproducing, below 0°C, and Others are found at the boiling temperatures of their aqueous environments. The crucial factor in both cases appears to be the availability of liquid water. In addition, organisms found at high temperatures must possess thermostable enzymes, which the majority of organisms do not have. Thus even though hot springs are unsuitable to support most forms of life, for some species they are a normal habitat. An environment is extreme only for organisms that are not adapted to it.

Organisms may survive if one factor is extreme but not if several are extreme. For example, a combination of low pH and high temperature is lethal for almost all forms of life. Biological molecules are subject to hydrolysis under these conditions, and cell death results. Additional environmental factors could be mentioned that, when high or low, can make the environment intolerable. The concentration of oxygen markedly affects the complement of enzymes that an organism may use for metabolism. Some cannot tolerate its presence and others cannot survive without it, and there are all shades in between. The level of essential nutrients can also contribute to environmental extremes. In deep aquatic environments growth is commonly substrate or nutrient limited. An organism requiring a complex diet would not do well in abyssal waters.

Air

Air is not a natural habitat for any type of individual cell, but it is for a wide variety of terrestrial multicellular systems. Unicellular organisms especially procaryotes like bacteria, derive some benefit from the air in being transported about by all currents to new and more favourable environments. But the major impact of air on the cell, if it is the only extracellular environment, is harsh and destructive. The most harmful effect of air on a cell is to brings its water content down to levels at which metabolism is impossible. Very few unicellular organisms can combat the determined forces of desiccation or a decrease in water activity.

Structural modifications are frequently employed by cells in response to this environmental problem. One is the construction of a cell wall that cuts down on evaporation. As an example, the bacterium that causes tuberculosis, *Mycobacterium tuberculosis, is* surrounded by a cell wall rich in waxy lipids that permits survival for months in dried sputum or dust. This wall can be viewed as a structural adaptation that buys survival time in the air while the organism is transported from person to person. A second type of modification brings certain cells into another form of dormancy. In bacteria, a structure called an endospore is generalized. Its thick coat and unique composition create an environment within which the cell's DNA and a few critical enzymes are held in it dormant state for an indefinite period. Another modification is demonstrated by eucaryotic cells, such as protozoa. They become encysted when under adverse environmental conditions by a protective wall. Cells that are thus protected can withstand the drying effects of air for long periods. When the environment is appropriate, a new vegetative cell emerges from the cyst.

Radiation

Sterilizing factor in sunlight is the ultraviolet component of the radiation that is selectivity absorbed by nucleic acids. *Mycobacterium tuberculosis* the same cell that can avoid desiccation for months, cannot survive in sunlight for more than a couple of hours. The most important detrimental effect is the photodimerization of thymine to form a cyclobutane dimer. This blocks the replication of DNA, accounting for the lethal and mutagenic effect of ultraviolet radiation on organisms.

Apparently destruction by this means is significant enough that mechanisms have been developed combat it.

Bacteria possess very low levels of an enzyme called photolyase. It absorbs light maximally at 380 nm, which is in the visible range. Light at this wavelength activities the enzyme to reverse the dimerization reaction and regenerate normal independent thymine residues. This process, termed *photoreactivation,* permits bacteria to survive better in spite of the destructive effects of ultraviolet light. Bacteria, other unicellular organisms, and multicellular organisms made up of small numbers of cells are most susceptible to destruction by ultraviolet radiation. Some of these employ screens that protect them. Pigment formation is an example of an antiradiation screen. Higher forms of life use a variety of protective screens, ranging from layers of outer skin cells to hair, feathers, shells, and exoskeletons. Obviously these structures have other functions too, but for the present discussion it is important to reflect on how cells have taken on special protective structures to deal with the life-threatening effects of their habitat.

Hydrostatic pressures

Living forms are found in the deepest parts of the oceans where the high hydrostatic pressure and absence of light have demanded extraordinary cell creativity for the maintenance of life. The deepest part of the ocean reaches down to 10,860 m at the bottom of the Challenger Deep in the North Pacific. Here the hydrostatic pressure is 1160 atm, but the region is not devoid of life. Both unicellular and multicellular organisms are present. Photosynthetic organisms are of course absent, being confined to the uppermost 100 m of the ocean through which light can penetrate at sufficient energies to activate photosynthetic machinery. The thicker the layer of water, the more the cell must make adjustments at the molecular level so that its metabolism is not blocked. Barotolerant and barophilic organisms, those tolerating and preferring high hydrostatic pressures, have adapted to this extracellular environment by using enzymes that are not adversely depressed by these conditions.

In general terms, hydrostatic pressure has a threefold effect oil biological systems: it inactivates enzymes, it affects physiological reaction rates and it influences growth or reproduction. Hydrostatic pressures of sufficient magnitude, on the order of 1000 to 3000 atm, denature many enzymes, whereas pressures lower than this cause enzyme inactivation.

It is therefore apparent that organisms that operate successfully at these pressures must have isoenzymes that are not seriously affected by the pressures. One Sspeculation is that barotolerant enzymes tend to be monomeric or dimeric, and their terrestrial counterparts polymeric. This is in line with the observation that high pressures depolymerize complex biological Structures. Protein synthesis represents the result of the combined action of many enzymes, so although the data do not pinpoint the molecular location of the pressure effect the impact on this obligatory cell process is apparent.

Whereas most enzyme activities are depressed at high pressures, certain enzymes are pressure activated. This is true for the enzyme fructose diphosphatase, an important enzyme in gluconeogenesis. The enzyme, in this case, does not appear to exist in a markedly special form for barophilic organisms. Both the surface-dwelling trout and the abylssal rat-tail possess quite similar pressure activated fructose diphosphatases. Certain kinetic parameters of the enzymes reveal differences that provide selective advantages for the environments in which these two organisms are found. In contrast to fructose diphosphatase, pyruvate kinase is inhibited by pressure and lactate dehydrogenase is pressure independent. Other enzymes could be placed into these three categories as well.

On sol-gel transitions within cells, hydrostatic pressure has an important role. At the molecular level, these transitions are the expression of a change from monomer to polymer forms of proteins respectively. The formation of hydrogen bonds is encouraged at increased pressures and ionic and hydrophobic interactions are discouraged. Therefore, cellular processes such as cell division and amoeboid movement that depend on the formation of microtubules are inhibited at high pressures. At pressures greater than 400 atm, dividing sea urchin eggs revert to a one-cell stage and pseudopodia formation in *Amoeba proteus* is inhibited.

Effect of high pressure on cell membranes has been studied by observing changes in cell shape and movement. The rationale for this is that the cell membrane is a contributing factor to cell shape and thus an effect on the membrane is transmitted to that of a shape change. Cell movement is observed, for example, in amoebae is also a membrane-related event.

Therefore a change in properties of the membrane may indeed be expected to alter movement. Change in both shape and movement are thought to be related to the sol-gel transitions of the cell that are pressure sensitive. Sol formation is preferred at higher pressures. Therefore a loss in shape or cell movement is reasonably attributable to sol formation in the protoplasm near the cell surface where, under normal circumstances, polymerized proteins make up a scaffolding structure governing membrane movements.

At pressures between 300 and 400 atm, amoebae become spherical and quiescent. The effect is reversible within a few minutes when the organisms are brought back to normal pressures, suggesting sol-gel transitions as likely candidates for the molecular target of this effect. Pinocytosis, a membrane function, is also blocked at increased hydrostatic pressures. High pressures cause both ciliates and flagellates to round Lip and lose their locomotive abilities. Flagellates are somewhat more resistant to pressure-induced shape changes than ciliates, possibly because of the presence of the pellicle, a specialized surface structure possessed by these organisms. High pressures generally cause a loss of flagella as well, and thus an inhibition of motility.

Interactions between cellular extracellular environment

The full impact of the environment on the cell is still somewhat beyond our reach, but some principles of interaction between the two arc becoming clearer. It is clear that cells, do not tolerate an extracellular environment of air and must be protected from it. The interaction is thus one primarily of defense. The desiccating effects of air and the radiations of the atmosphere have severely limited the habitat for unicellular forms of life.

These environmental factors have encouraged the generation of multicellular systems in which the exterior cells take on the role of protection for the rest. Cell wall structures have also been promoted in some organisms to protect them from the harshness of the air. As vital as water is to life, it is also clear that environmental water can pose a problem for cells. An environment of fresh water requires defensive tactics to keep cells from bursting as a result of high internal osmotic pressures. The defense most frequently used is the cell wall, which is able to sustain extremely high internal pressures. In exchange for this protection, the cell has sacrificed flexibility and often freedom of movement.

Water at high pressures has given the cell another set of problems. The interaction in this case is between water and proteins, both those that have a structural role in the membrane and in microtubules and those that have enzymatic functions. Protein conformation is changed at high hydrostatic pressures, depressing catalytic activities and inhibiting polymerization reactions. Some cells have successfully adapted to these conditions by developing and using proteins that are less sensitive to pressure denaturation and in some cases are even activated enzymatically by high pressures.

Salinity of seawater is usually significantly higher than that of cells and is of such a composition that cells must make special compensations to survive. The effect of this extracellular environment is thus both dehydration and enzyme inhibition. This has required cells to modify their membranes so that water can be sequestered From the environment, and halotolerant isoenzymes are employed. This is yet another defensive ploy ploy to cope with the environment.

Cells have taken offensive action toward their environments by the creation of their own extracellular matrices. Synthesis of matrix not only facilitates intercellular adherence but creates a cell Community wherein cells are assigned specialized tasks. In multicellular organisms the destructive effects of air are thwarted by surface cells that are antidesiccant and able to protect against radiation damage. The aqueous environment in which the cells are bathed is cell generated with optimal levels of ions of an appropriate composition. As long as there is an external source of vital materials, the extracellular environment is maintained.

Chapter 3

Foundations of Genetics

This chapter briefly describes the biological basis of heredity, how the process is regulated, how genetic variation arises, and how this genetic variation is detected using molecular genetic-techniques. The biological basis of heredity includes the fact that genes are contained on structures called chromosomes. The linkage of genes that lie close together on chromosomes has made possible the mapping of the human genome. Moreover, abnormalities in chromosomes contribute importantly to behavioural disorders, especially mental retardation.

Mendalian studies

The person who laid the scientific foundations of genetics was an Austrian, monk, Gregor Mendel. Mendel, trained in mathematics to begin with, began in 1857 experiments with garden peas, the common edible type. Mendel succeeded where all others failed because of three main methodological innovations. First, Mendel did not attempt to see whether offspring were generally like or not like their parents. Instead he studied the inheritance of unit characters. He selected some particular feature in which offspring and parents resembled or differed from each other and followed this regardless of other likenesses or differences. Second, Mendel used quantitative methods. He did not merely list like and unlike, but counted how many were like one parent and how many were like the other. Third, Mendel set up a theoretical scheme based on probable reproductive processes that he felt explained the results. This scheme then suggested further tests.

Mendel's work, published in 1866, remained relatively unnoticed in the scientific world. A few scientists mentioned it in books or articles, but failed to see the real significance of his results. Most either ignored it or seemingly failed to understand what Mendel was driving at. When Mendel died, his results were still more or less hidden in the journal in which he published them. Part of the lack of appreciation of Mendel's work was probably due to the fact that Mendel developed theoretical ideas that he could not support with actual observations. The observations that supported his vision were made later, between 1870 and 1895. For his experiments Mendel selected the common garden pea.-He did this for two reasons: many varieties which are true breeding are easily available, and peas have flowers that are self-fertilizing, so that one knows the source of the pollen (male) that fertilizes the egg (female).

Mendel selected seven pairs of contrasting appearances for study. Each of these he studied first independently, then in combination with the others. Let us take as an example his crosses of tall, or climbing, plants with dwarf, or bush, plants. This kind of mating is called a *monohybrid cross*. To do this he removed the stamens from the flowers of one of the plants, leaving it an artificial female, then transferred pollen to this from another plant. Almost everyone at that time would have expected that such a cross would produce intermediate plants—that is, neither tall nor dwarf, but of in-between height. Instead, when he planted the peas the following year they all grew into tall plants. It was as if the dwarf character had disappeared. Mendel now allowed these tall plants to self-fertilize. The resulting peas when planted produced 3/4 tall plants and 1/4 dwarf. The dwarf feature obviously had not been lost. Mendel now knew that there were two types of tall plants, one like the original parent plant that produced all tall offspring, and the other like the hybrid plants that produced partly tall and partly dwarf offspring. So he let the plants of the mixed group fertilize themselves. All the dwarf plants produced only dwarf plants. The tall plants, however, could be divided into two classes. Subdividing the 3/4, he found that 1/4 produced all tall plants (as the 1/4 dwarf plants produced all dwarf plants), and 2/4 produced a mixture of tall and dwarf in the same ratio as had the parents, i.e., 3/4 tall and 1/4 dwarf.

Mendel found the same situation for each of the seven characters he studied: individuals in the first generation after a cross all were like one parent; these, when self-fertilized, produced offspring in the ratio of three like the original parent whose character appeared in the first generation and one like the other original parent. The next step was to ask what could give a ratio of 25 percent A, 50 percent A′ (that is, appearing like A, but not like it in potentialities), and. 25 percent B. Being mathematically inclined, Mendel saw that one got such a ratio if he expanded the binomial $(x + y)^2$, which gives $1xx + 2xy + 1yy$. One might just as well say that he would get the same ratio if he were to toss two similar coins simultaneously. The chance for either coin to be head or tail is 50 percent. The chance that both coins could come up heads simultaneously is, therefore, 50 percent x 50 percent (.5 x.5) or 25 percent. There is the same probability for two tails. Since there are two ways in which one can get a head and tail the probability is twice 25 percent (.5 x.5 x 2).

To Mendel, this meant that each tall plant produced by mating a tall with a dwarf plant must have received from each of its parents one determiner for tallness and one for dwarfness. When present together, they resulted in a tall plant When this plant produced its reproductive cells, each reproductive cell received either the determiner for tallness or that for dwarfness. When these recombined in fertilization there was only one way in which the two determiners for tallness could be joined, and similarly for dwarfness, but there were two ways in which the deter miners for tallness and dwarfness could be joined. Those plants that received either one or two of the determiners for tallness were tall, while the others were dwarf.

Mendel found that it did not matter whether he took pollen from a tall plant to put on the eggs from a dwarf plant or vice versa. He had shown that both parents contribute equally to the inheritance of the offspring. Mendel had originally set out with the hypothesis that crosses would produce intermediates, and in testing he had not confirmed this. He had to revise the accepted generalizations. So he developed a new theory: that contrasting appearances of the same character are determined by individual determiners that are passed on through the reproductive cells to the offspring. These determiners, he postulated, are present double in adult organisms, but occur singly in gametes. Combination of gametes restores the double number and allows the development of a new individual. He found that when two different determiners are present, the dominant, appears as if the other were not present at all, and the other does not appear.

In the case of the cross of the tall and dwarf plants, Mendel found that the offspring were actually a little taller than the parent plants. He had discovered, without realizing it, what we call today hybrid vigor, resulting from the cross of two lines of plants that have for a long time been inbred. This discovery, remade much later, became the basis for the development of high-yield hybrid corn and other plants. This, however, was far in the future. Mendel now proceeded to test the theory by setting up predictions based on the theory and then doing experiments to see whether they were borne out in practice. Thus he could predict that if one were to cross an F, tall plant (heterozygous) with a P, tall plant (homozygous), the offspring would all be tall, but one half of these, when self-fertilized, should yield a 3: 1 ratio. This -roved to he the case. Today we call this a *backcross to the dominant,* for it is as if one crossed the individual back to one of its parents.

Next, he could predict that if one were to cross an F_1 tall plant with a P, dwarf plant, he should get 1/2 tall and 1/2 dwarf plants (1:1 ratio). This indeed was the result. This is now called a *backcross* to the recessive, or, because it can be used as a test for heterozygosity of an individual showing a dominant character, it is called a *testcross*. The results of these crosses therefore proved to be as predicted, and so his theory was supported. When one considers four, five, or six characters simultaneously, the ratios become very complicated indeed. If one were to test these he would need large numbers in the populations to reveal all of them.

For instance, even with only two pairs of characters the absolute minimum number of offspring that would have to be counted would be 16, and if one counted only 16, he would have a very, very high probability that some classes would appear more than expected while others were absent. If be were to have any reasonable hope of determining the ratio, he would have to count well over a couple of hundred individuals. And this is for two sets of characters only. For three characters the minimum number of individuals would be 64, and as before this is a totally unrealistic minimum. Probably at least a thousand or more would be needed to be sure to get reasonable approximations of the ratios.

So, as Mendel saw, with plants and animals that did not produce large numbers of offspring, anyone attempting to find any sensible relationship while dealing with a number of characters is certain to be frustrated.Mendel was able, by his analytical methods, to lay the foundations of genetics, foundations that have stood the test of time. Basically he postulated that there are two determiners, now called genes, for each individual character in the adult organism. These two, he postulated, separate at the formation of the reproductive cells, so that an individual passes on only one of the pair to each of the offspring, the other gene in the offspring coming from the other parent. If the genes are alike, the appropriate trait develops. If they are different, one appears in the offspring, i.e., is dominant, while the other disappears for the moment, i.e., is recessive.

There is, however, no blending of genetic materials, as many of Mendel's predecessors had thought. A recessive gene is unaffected by being present in combination with a dominant gene, and produces its effect whenever it join with a similar gene. Two sets of characters in the same plant are determined independently of each other, each gene pair acting as if the other were not present. Mendel's work went unnoticed by almost all biologists of his day, probably because he seemed to be making too many assumptions that seemed unjustified to them, and because he used mathematical models, which at that time were unfamiliar to biologists. Mendel, it must be admitted, had gone out on a limb, so to speak, in postulating—without definite evidence—that gametes contain only half the genetic material of adult cells. This was an explanation based upon pure imagination, but seemed to Mendel inescapable on the basis of his results. The evidence that was to make this plausible was to come some years later.

Shortly after Mendel's death the German scientist Weismann, studying how variations arise and are transmitted in animals, performed a series of experiments that showed that acquired characters could not be inherited. He became convinced that the germinal material—that is, the material that is passed on from generation to generation-is unaffected by what happens to the individual that houses it. In each generation the zygote produces many cells, the somatic cells, that form the body and enclose the reproductive *germ cells.*

However, according to Weismann, early in development the germ cells are set aside and remain unaffected by what happens to the somatic cells. This idea was supported by his discovery of a type of embryonic development in which at a very early stage, sometimes at the four- or eight-cell stage, certain cells are set aside and do not divide further until they produce gametes in the adult. This was considered as strong evidence that gametes-to-be are set aside early in development and that the development of somatic cells goes on independently, merely producing a house and transferral structures for the germ cells. Weismann pointed out that the somatic cells die, but the germ line is potentially immortal. This idea emphasizes the importance of the germ line and the fact that body cell produced from it do not feed back information to the germ cells. It supports Mendel's concept that genes are not affected by being present in organisms that have features unlike those that the genes would produce. Even more important in giving support to Mendel's ideas was the discovery of the mechanism of formation of gametes and fertilization. This was made possible by the discovery of mitosis about 1875. In mitosis the chromosomes are evenly divided, so that daughter cells contain the same number of chromosomes as do parent cells.

On theoretical grounds alone it is obvious that, were gametes to be produced by mitosis, fusion of gametes would double the chromosome number. Unless fertilization were followed by fusion of chromosomes which is theoretically possible, but just does not happen the chromosomes would have their own population explosion. A series of observations on development of embryos revealed the true situation. We can here mention only the work of Boveri, who was one of many biologists studying embryology at the end of the nineteenth century. He studied the embryonic development of the horse nematode, *Ascaris* (now *Parascaris*) *megalocephala*. Boveri's interest was in embryology, and he used this animal because the eggs develop as they move along the uterus of the worm, so that, by preserving the uteri and making longitudinal sections, he could follow the time sequence of embryology. He was thus able to study fertilization, and since Parascaris has very few chromosomes, he could clearly see what happened to them.

Boveri found that, while the adult animal or developing embryo has four chromosomes, an egg or sperm has only two. Fertilization was a means for restoring chromosome number by fusion of gametes that each have only half the chromosomes of body cells. Further work by man, investigators with many animals confirmed this. The discovery, about 1895, of *meiosis*, by which gametes are produced, Proved that gametes receive only one member of each pair of chromosomes present in adult cells. So an individual animal receives from each parent one member of each pair of chromosomes characteristic of the species. It takes little imagination, with this knowledge, to see the mechanism by which Mendel's results could be explained, if one assumes that each chromosome acted like a gene.

A moment's reflection, however, will show that this is an untenable proposition, for an animal such as Parascaris, with only four chromosomes, could hardly be thought to have only two differentiable characters in development. Or an animal such as man, with 23 pairs of chromosomes, could hardly, be thought to have only 23 separable developmental potentialities. It must be obvious that, if one is to assume that chromosomes are involved in the separation and reunion postulated by Mendel, the chromosomes must consist of many genes. Individual females can produce large numbers, up to some hundreds, of offspring, so that genetic ratios needing large numbers can be tested.

When Morgan and his students started, the *Drosophila* that they reared were all pretty much alike, a wild population. However, as they proceeded with their program of what we now call inbreeding—i.e., crossing individuals from the same parents—which increases the chance of turning up recessives, they discovered individuals that were different from those in the normal population, called technically mutants. The mutant types were crossed with wild flies, and the F_1 insects were then mated together to produce homozygous stocks of the heritable variations. There was no one else, as far as we know, who was working at the time that Mendel was with the same methods and insights that Mendel used. His ideas were so radical that his contemporaries were unable to understand them.

When Mendel sent a report of his results to Nageli, a botanist, Nageli found them incomprehensible, even unbelievable. He did not see the true significance of Mendel's work, but then neither did anyone else. It is beside the point to say that Mendel himself did not carry his idea to their full conclusion. He certainly was not encouraged to do so.

Mendel was truly ahead of his time, and his was truly a personal achievement. Just as the symphonies of Beethoven or the plays of Shakespeare represent the distillation of the experience and creativity of individuals, so the concepts of Mendel—which have a beauty like that of a symphony or work of art—represent an individual achievement of genius. Scientific ideas such as those of Newton, Einstein, or Mendel are truly products of a creative human mind, for only facts exist in the outside world. As with art and music, one generation of workers in science stands upon the shoulders of the preceding. When Mendel started his work, he probably expected to get intermediate conditions in the offspring of his original crosses. When crossing a tall pea plant with a dwarf plant, he expected to get intermediate-height plants, because the general belief was that the characters from the parents blended in the offspring. An analogous situation would be the mixing of red paint with white paint to get pink paint.

Mendel, however, found that his mixing, figuratively speaking, produced red paint. One feature was dominant over the other With the extension of studies in genetics, however, it soon became obvious that this situation was not always the case. For instance, in crossing a red-flowered four-o'clock plant with a white-flowered one, investigators produced offspring that had pink flowers. This seemed to be a situation somewhat as Mendel might have expected. However, when the pink-flowered plants were self-fertilized or crossed with each other, the offspring appeared in the ratio of 1/4 red, 1/2 pink, and 1/4 white. So the red and white appeared again. Obviously, therefore, there was no blending, no intimate mixing of red and white. They remained distinct. This was at first called incomplete dominance; however, this is somewhat of a misnomer. Today we call it *intermediate expression,* in which the heterozygote has a phenotype intermediate between that of the homozygotes.

The genetic integrity of the genes is not lost. Mendel's laws apply as well here as to the situations that Mendel studied. It is only the expression of the genes when present in the heterozygous condition that differs. We now know many cases of this type, particularly in the inheritance of what can be called quantitative characters, i.e., height, degree of pigmentation, and so on. The genes are known as *quantitative genes,* for it is as if each gene adds a certain amount of the character that it determines. In most cases we now know a series of gene pain, determines a particular character, with one set of alleles producing one extreme, and the other set the other extreme. This may be illustrated by a simplified picture of skin colour inheritance in man. It is assumed here that the genes, +, contribute dark pigmentation to the skin, each gene contributing a particular quota, the alleles, *s*, of these contributing quotas of whiteness.

We know that at least three gene pairs are involved-almost certainly more-but we assume for our illustration this simple situation, The cross of two triple homozygotes would produce, as shown, a triple heterozygote. This individual would therefore have an intermediate skin colour. If two of these triple heterozygotes were to mate, it is possible to get all shades of colour. It is obvious, however, that the triple homozygotes, restoring the grandparental colours, would appear only in 1/64 of the cases. Most individuals would have intermediate skin pigmentation. This type of quantitative inheritance is found in many cases of size, length, height, etc. One does not in these cases, therefore, get a neat 3: 1 ratio that Mendel obtained.

Most of the complex features of an animal or plant body are not inherited merely through single genes. So gene interactions among series of genes acting upon some particular character are usual. Many of these genetic galaxies were discovered a few years after the discovery of Mendel's work, and again seemed at first to be contradictory. We shall discuss only one case to show how the seeming difficulty is easily resolved. In mice, one can clearly differentiate albinos—with no fur colour and with pink eyes—from coloured mice.

The typical Mendelian type of inheritance is shown also by a backeross of an F, coloured mouse to the recessive albino, which results in 1/2 coloured and 1/2 albino. However, all mice are not either albino or coloured like wild mice. They may have many different coat colours. For instance, a number of strains of mice have black fur.

Assuming a simple situation, if one crosses an albino mouse with a black mouse, he obtains in the F1 all wild-appearing (gray) mice. The albino has two genes for albinism (*a/a*), while the black has two genes for colour (+/+), meaning any colour. The albino, in our example, is assumed to be carrying a separate pair of genes for wild coat colour (+,/+). which are dominant over those for the black colour *(b/b)* carried by the black mouse. The F_1 mice, therefore, have the recessive gene for albinism with the dominant for colour and all will be coloured; they also have the dominant gene for wild colour with the recessive for black, so are gray in colour. When the F_1 mice are inbred, the results, are as shown, a 9:3:4 ratio. This may not appear to be what one would expect in Mendelian inheritance, but it is fully explicable, as shown, in Mendelian terms. The coat colours of mice, rabbits, rats, cattle, dogs, and other mammals have been studied in detail, and complex gene interactions of this sort occur in all.

Sexual inheritance

The mechanism of sex determination obviously has some genetic bases. Studies between 1895 and 1905 on various species of insects showed that each species has a characteristic number of chromosomes which appear in pairs-pairs, that is, provided one examines chromosomes from a female. If, on the other hand, one examines those from a male, all the chromosome pairs match those of the female except for one pair. In this pair one chromosome, named the X chromosome, matches the corresponding set of the female, but the other, named the Y chromosome, does not. In some insects the male has only the X chromosome in this set, but has no mate for it-the male has an uneven number of chromosomes.

When eggs form, each receives one member of each pair of chromosomes, and so has n + X chromosomes. When sperms form, each receives one member of each of the pairs that are common to both sexes, named, autosomes, and either an X chromosome or the unmatched one, the Y chromosome. If a sperm with n autosomes and an X fuses with an egg—all eggs having n autosomes and an X—the combination $2n$ + XX results, and that produces a female. If a sperm with n autosomes and a Y chromosome fuses with an egg, $2n + XY$ results, and that produces a male. Since the two classes of sperms are produced in equal numbers, the sex ratio is 1: 1.

The X and Y chromosomes are, therefore, known as sex chromosomes, because their distribution in the zygote determines the ultimate sex of the individual produced. This type of sex determination-in which the female has two similar chromosomes (XX) and the male either has one of these and another that is different (XY) or has an unpaired X (XO)—was the first type discovered. Many insects have XY type of sex determination, as do many mammals. Thus in man the female has 22 pairs of autosomes plus 2 X chromosomes, while the male has 22 pairs of autosomes plus X and Y chromosomes.

In birds and some other animals the male has the matched pair of sex chromosomes, while the female has the unmatched set. This is often called a WZ type of sex determination, the male being WW, the female WZ. Sex, therefore, is determined in most animals at the time of fusion of egg and sperm. In the case of man, the female produces only one type of egg. The male produces two types of sperms in equal numbers. Fusion of a sperm with an egg is by chance, so approximately 1/2 males and 1/2 females are produced.

Actually, in man, a slightly larger proportion of males is conceived than females. This has been attributed by some biologists to the fact that the X chromosome is larger than is the Y, and therefore the X-bearing sperm is somewhat larger than the Y-bearing sperm. The latter, therefore, has a very slight edge in the race for the egg. Statistically this would result in a slight excess of male zygotes over female zygotes, as is found. There is indeed a difference between X-and Y-bearing sperms, such that separation might be possible. If so, appropriate treatment of the semen may allow the X-bearing sperms to be separated from the Y-bearing sperms.

Man could thus determine the sex of offspring. This has obvious applications in animal husbandry. It also has implications in human societies. In some countries a man is allowed to divorce his wife if their union produces no sons. Obviously the wife has no control over this. The male is no more responsible, as far as knowledge is concerned, but it is the sperms that differ. We are on the threshold or control of sex through knowledge and at the same time find societies that have not even absorbed the elements of that knowledge. With *Drosophila* a mutant was found that seemed to have a different pattern of inheritance.

In Mendel's work no difference in results occurred if pollen came from one plant or from another-reciprocal crosses gave the same results. In the earlier work on coat colours and similar features of mammals and on the first mutants discovered in *Drosophila,* the same was true. Then a mutation producing white colour in the eye instead of the usual red was found. This gave somewhat different results. Females of the F_1 appear as one would expect from Mendelian rules. Males, however, from matings; of white-eyed females with red-eyed males do not. The males seem to inherit their eye colour from the females, in spite of the supposed dominant gene for red eyes from the male. Furthermore, F_1 males when bred further seem to be homozygous for the eye colour, while F_1 females seem to be heterozygous. This puzzling situation can be explained if we assume that the genes for white and red eye colour are on the X chromosome, and that the Y chromosome has no genes on it. The latter assumption gets support from the observation that, in some insects, the Y chromosome may be completely absent, and in all it is usually much smaller than the X.

Since the males receive their Y chromosome from their father—otherwise they would not be males—they can receive a gene for eye colour only from their mother on the X chromosome and therefore will have eyes like hers. The females receive an X chromosome from each parent, and so have the situation the same as for ordinary inheritance. This phenomenon is called sex linkage. It is as if the inheritance of a gene depends upon the sex of the individual. In a sense it does, but only because sex itself depends upon the presence of X and Y chromosomes with their genes. In XY or XO sex determination males always show sex-linked traits as if the genes were dominant.

Males cannot be heterozygous. Females, having two genes for sex-linked traits, inherit sex-linked characters just as they do autosomal traits. Mendel could not have discovered this, for he did not study an organism that had sex determination like this. The results, however, can be interpreted by Mendelian concepts, with the addition of the postulate that one chromosome of a pair may not bear genes. The white-eye gene is famous in the history of genetics, for it afforded investigators an experimental test of the chromosome and gene theories of heredity.

In man some important traits are sex-linked—e.g., red-green colour blindness is inherited as a recessive, sex-linked gene. Sons seem to inherit the trait from their mothers, while daughters do not necessarily seem to inherit from their fathers. Females can be heterozygous, that is, carriers, for colour blindness while having normal vision themselves. Males cannot be, for they have no gene to "cover up" the recessive if they get it on the X chromosome. Females with normal vision, therefore, if they are carriers, will have sons who are colour blind in equal numbers with those that have normal vision. A much more serious trait that is inherited as a sex-linked recessive is hemophilia, a defect in which the blood fails to clot. Mendel's original discoveries were made with seven sets of contrasting phenotypes in peas. Each of these was inherited seemingly independently. When it was known that the determiners of phenotypic characters are genes that are part of chromosomes, it became obvious that sooner or later genes would be found that would be on the same pair of chromosomes.

Drosophila, for instance, has only four pairs of chromosomes, and it is obvious that the determination of the total body pattern of an animal as complex as this would require many more than four pairs of genes. If two pairs of genes were on the same pair of chromosomes and remained thus, one would expect them to travel together in matings. In short the two pairs of genes would act as if they were one, and one might be at a loss to decide whether this was the case or whether two pairs of genes were present.

In his studies Mendel used seven pairs of characters. We know today that the garden pea has seven pairs of chromosomes, Somehow Mendel had selected seven pairs of genes, each on a different pair of chromosomes. So the characters were inherited independently. Had Mendel studied one more pair of genes these would have been on the same pair of chromosomes as one of those he did study, and the results would have been quite different. Perhaps he might have understood what was going on, but this seems unlikely, for chromosomes were unknown in his day. With *Drosophila,* mutants were soon found whose genes were on the same pair of chromosomes-they were said to be *linked.* We can take, as an example, genes that determine wing length and body colour.

If we were to cross a fly that had vestigial wings and was black in colour with a wild—i.e., winged, tan—fly, we would expect the F_1 to be all wild. If the genes remained fixed on their chromosomes—i.e., showed total linkage—we would expect the F2 to be 3/4 wild and 1/4 vestigial-black. This, however, is not the case. Instead the F_2 flies include not only wild and vestigial-black, but also some with vestigial wings and wild (tan) body and a corresponding number with wild-type wings and black body. It is as if the genes for vestigial and black become separated, as if one set of genes exchanges places on their chromosomes. This phenomenon is called *crossing-over.* Crossing-over does not happen enough to make the genes act as if they are independent of each other. Instead it occurs in small proportions, but in fixed percentages with given sets of genes. It soon became obvious that the percentage of crossing-over between genes on the same chromosome is a function of the distance between them.

When chromosomes pair in the premeiotic stage called *synapsis,* they twist around each other. At this time they are already double, preliminary to the double division of meiosis. So they form four-stranded, twisted structures called tetrads. Apparently while so twisted, the chromosomal strands may break where they cross, and instead of reforming the original chromosomes, re-attach to the other strand. We can show this, in simplified fashion, by assuming a two-strand condition, each strand representing one member of the pair of chromosomes. If a break occurs between the two pairs of genes that are on the same chromosome and the chromosome pieces exchange partners, the genes apparently exchange places.

Crossing-over is the result of such breakage and reunion in chromosomes, as repeatedly shown by experiments. Obviously in such a case the greater the distance between the genes on the chromosome, the greater the chance for crossing-over to occur: if two genes are close together the chance for a break to occur between them is small; if they are far apart the chance for breaks is large. So the percentage of crossing-over between any two genes is a measure of the distance between them. Crossing-over can therefore be used to determine the order of genes on chromosomes and to give a quantitative measure of their separation.

In short, by studies on linkage and crossing-over one can produce *chromosome maps.* To illustrate, if one has three mutant genes, a, b, and c, all on the same chromosome, he could mate an animal with this chromosome *(abc/abc)* with a wild animal (+++/+++), producing a situation in which *a, b,* and *c* are on one member of the pair of chromosomes and the wild alleles on the other. The triple heterozygotes can then be backerossed to triple recessives, as a testeross. The testcross maximizes the proportion of visible cross-over lasses in the offspring. If one finds that there is 15 percent crossing-over between a and b, and 10 percent between *b* and *c*, we need only determine the level of crossing-over between *a* and c to know whether the order is *a-b-c, a-c-b, or* some other. If it is *a-b-c* the percentage of crossing over for *a-c* would be approximately the sum of *a-b* and *b-c*. For *a-c-b*, crossing-over for *a-c* would be less than that for *a-b*. By thus proceeding along a chromosome, using genes whose positions are already known, one can determine not only the order but also the genetic distances between the genes.

We say genetic distances, for these are not necessarily physical distances. Chromosomes are not equally susceptible to twisting and breakage all along. This great triumph of genetic science led, particularly from about 1915 to 1930, to many studies mostly with *Drosophila* and corn, designed to discover new mutant genes and to locate them on chromosome maps. It gave an entirely new precision to the science of genetics, a type of precision that was uncommon in most biological science at that time.

Geneticists, however, still had to wait for mutations to turn up. They continued to inbreed fruit flies or other organisms, select variant animals or plants, and then, through crossing these with wild stocks and inbreeding again, try to establish pure lines of particular mutants. Mutation was obviously a rare process, and so progress was slow.

Then, in 1927, H. J. Muller, discovered that the ordinarily very slow rate of mutation in *Drosophila* could be increased by irradiating the organisms with x-rays. The increase, he found, was proportional to the dosage—that is, the intensity and time of application. Most of the mutations that were produced by x-ray techniques proved to be so deviant from normal that they were *lethal*—that is, they killed the individuals receiving them.

Since in many cases death occurred before the individual became an adult or even before hatching of the egg, their occurrence was difficult to establish.

So a special technique, the *CIB* method, was developed to test for lethal mutations. Using this technique, which enables one to detect mutations that kill an animal even early in life, Muller and fellow geneticists studied in detail the mechanisms of mutation. Since then it has been found that ionizing radiations produced by radioactive materials also bring about increases in mutation rates. This discovery is the basis of the debates over possible genetic damage to man resulting from irradiation in atomic bombing, bomb tests, or even peaceful uses of atomic energy. It is certain that any increase in radiation dosage, whether from radioactive materials or from x-rays, increases the rate of mutation in man as well as in other animals. The rate of increase may seem to be small and the mutations are not likely to appear phenotypically immediately, for most of them will be recessive and therefore covered *UP* by dominant alleles with normal effects.

However, over many generations, particularly if there is appreciable inbreeding, the mutations will express themselves. The point for debate, therefore, is how much greater load of harmful mutants the human race can sustain without excess strain upon the economic and social fabric resulting from appearance of abnormal individuals. This matter is not easy to settle, and even biologists disagree badly about the matter. Shortly after Muller's original work another important discovery was made. Chromosome maps were constructed without geneticists being able to relate them to specific chromosome structures.

The chromosomes of the cells of most animals, including *Drosophila,* are very tiny, and ordinary microscopes could only distinguish that generally chromosomes are not uniform in appearance all along. There was no way, however, to relate an exact spot on a chromosome with a particular gene. Then anothe scientist discovered that the chromosomes in the salivary glands of the larvae of *Drosophila,* the now famous salivary chromosomes, are really giants. Furthermore, these chromosomes are in a permanent state of synapsis, so that the two members of the chromosome pair are twisted together. Salivary chromosomes of *Drosophila* are made visible by crushing the salivary glands of a larva on a microscope slide and staining them. The bands which show so plainly on these chromosomes allow one to map the chromosomes morphologically.

The problem is, however, to relate to this morphological map the chromosome map derived from linkage and crossing-over studies. This has not been satisfactorily done for a number of reasons which we shall not go into. However, a great step was taken when it was discovered that some mutations are not chemical changes in the chromosome and so invisible, but are changes in the gross structure of chromosomes. These changes are particularly likely to occur with x-ray treatments, which in many cases cause breakage of chromosomes. The chromosome fragments thus detached may re-attach to another chromosome; these are translocations. In other cases the fragments seem to be spun around and re-attach to the same chromosome but are inverted, these are *inversions.* In still other cases the fragments are knocked out of the chromosomes entirely; these are *deletions.*

Deletions could obviously be detected by comparison with normal chromosomes. Inversions result in loops when the chromosome with an inverted section synapses with its normal partner. Translocations similarly result in complex chromosomal patterns, in which there seem to be fewer than the normal number, for the translocated piece synapses with the corresponding piece of the partner, thus joining two sets. So mutations that were the result of chromosomal alterations could be detected and mutant stocks bearing these mutations could be used to establish positions of genes on the chromosomes-or at least so it seemed.

Drosophila had proved to be an unexpectedly happy choice for gene studies. As Morgan knew when be selected it, it reproduces rapidly and bountifully. It is small enough to live easily in small spaces, with little cost for food. Besides these advantages, however, *Drosophila* proved to be among the few organisms for there are only certain other flies similarly endowed-having salivary chromosomes. These allow the geneticist to match results of quantitative studies on linkage with observable morphology of the chromosomes. At this point it seemed as if the concept that each chromosome is an aggregation of genes was beyond doubt.

All the evidence from crossing-over studies led to the conclusion that the chromosome is like a string of beads, the genes like beads linearly arranged on the string. Chromosomes were known to consist of proteins with other organic materials. Biologists believed that a chromosome was probably a set of protein strands on which the genes, of unknown chemical nature, were strung. It was expected that genes would prove to be organic compounds whose atomic arrangements could be altered to produce new compounds, mutant genes, resulting ultimately in mutations.

Human species has 23 pairs of chromosomes, for a total of 46 chromosomes. The number of chromosome pairs varies widely from species to species. Fruit flies have 4, mice have 20, dogs have 39, and carp have 52. Our chromosomes are very similar to those of the great apes (chimpanzee, gorilla, and orangutan). Although the great apes have 24 pairs, two of their short chromosomes have been fused to form one of our large chromosomes. One pair of our chromosomes is the *sex chromosomes* X and Y Females are XX and males are XY All the other chromosomes are called *autosomes*.

Chromosomes have characteristic banding patterns when stained with a particular chemical. The *bands*, whose function is not known are used to identify the chromosomes. At some point in each chromosome, there is a *centromere,* a region of the chromosome without genes, where the chromosome is attached to its new copy when cells reproduce. The short arm of the chromosome above the centromere is called *p* and the long arm below the centromere is called *q*. The location of genes is described in relation to the bands. For example, the gene for Huntington's disease is at 4*p*16, which means the short arm of chromosome 4 at a particular band, number 6 in region 1. In addition to providing the basis for gene mapping, chromosomes are important in behavioural genetics because mistakes in copying chromosomes during cell division affect behaviour.

There are two kinds of cell division. Normal cell division, called *mitosis*, occurs in all cells, called *somatic cells*, except the sex cells that produce eggs and sperm (*gametes*). In mitosis, each chromosome duplicates and divides to produce two identical cells. A special type of cell division called *meiosis* occurs in ovaries and testes to produce eggs and sperm, which have only one member of each chromosome pair. Each egg and sperm has 1 of over 8 million (2^{23}) possible combinations of the 23 pairs of chromosomes. Moreover, crossover (recombination) of members of each chromosome pair occur's about once per meiosis and creates even more genetic variability.

When a sperm fertilizes an egg to produce a zygote, one chromosome of each pair comes from the mother's egg and the other from the father's sperm, thereby reconstituting the full complement of 23 pairs of chromosomes. A common copying error for chromosomes is an uneven split of the pairs of chromosomes during meiosis, called nondisjunction.

The most common form of mental retardation, Down syndrome, is caused by nondisjunction of one of the smallest chromosomes, chromosome 21. Many other chromosomal problems occur, such as breaks in chromosomes that lead to inversion, deletion, duplication, and translocation. About half of all fertilized human eggs have a chromosomal abnormality. Most of these abnormalities result in early spontaneous abortions (miscarriages). At birth, about 1 in 250 babies have an obvious chromosomal abnormality. Although chromosomal abnormalities occur for all chromosomes, only fetuses with the least severe abnormalities survive to birth. Some of these babies die soon after they are born. For example, most babies with three chromosomes (trisomy) of chromosome 13 die in the first month and most of those with trisomy-18 die within the first year.

Other chromosomal abnormalities are less lethal but result in behavioural and physical problems. Missing a whole chromosome is lethal, except for the X and Y chromosomes. Having an entire extra chromosome is also lethal except for the smallest chromosomes and the X chromosome, which is one of the largest. The reason why the X chromosome is the exception is also the reason why half of all chromosomal abnormalities that exist in newborns involve the sex chromosomes. In females, one of the two X chromosomes is inactivated, in the sense that its genes are not transcribed. In males and females with extra X chromosomes, the extra X chromosomes also are inactivated. For this reason, even though X is a large chromosome with many genes, having an extra X in males or females or only one X in females is not lethal. The most common sex chromosome abnormalities are XXY (males with an extra X), XXX (females with an extra X), and XYY (males with an extra Y), each with an incidence of about 1 in 1000. The incidence of XO (females with just one X) is 1 in 3000 at birth, because the vast majority of such conceptuses abort.

Fuctions of DNA

Nearly a century after Mendel did his experiments, it became apparent that DNA (deoxyribonucleic acid) is the molecule responsible for heredity. James Watson and Francis Crick proposed a molecular structure for DNA that could explain how genes are replicated and how DNA codes for proteins.

The DNA molecule consists of two strands that are held apart by pairs of four bases: adenine, thymine, guanine, and cytosine. As a result of the structural properties of these bases, adenine always pairs with thymine and guanine always pairs with cytosine. The backbone of each strand consists of sugar and phosphate molecules. The strands coil around each other to form the famous double helix of DNA. The specific pairing of bases in these two-stranded molecules allows DNA to carry out its two functions: to replicate itself and to synthesize proteins.

Replication of DNA occurs during the process of cell division. The double helix of the DNA molecule unzips, separating the paired bases. The two strands unwind, and each strand attracts the appropriate bases to construct its complement. In this way, two complete double helices of DNA are created where there was previously only one. This process of replication is the essence of life that began billions of years ago when the first cells replicated themselves.

Another major function of DNA is to synthesize proteins according to the genetic information that resides in the particular sequence of bases. DNA codes for sequences of the 20 amino acids making up the thousands of specific enzymes and proteins that are the stuff of living organisms. Some amino acids are coded by as many as six codons. Three codons signal the end of a transcribed sequence. This same genetic code applies to all living organisms. Breaking this code was one of the great triumphs of molecular biology. The human set of DNA sequences (genome) consists of about 3 billion nucleotide bases, just counting one chromosome from each pair of chromosomes. The 3 billion bases contain about 100,000 genes, which range in size from about 1000 to 2 million base pairs. About a third of our genes are expressed only in the brain; these are likely to be most important for behaviour. Genes code for sequences of amino acids that form the thousands of proteins of which organisms are made. Proteins create the skeletal system, muscles, the endocrine system, the immune system, the digestive system, and, most important for behaviour, the nervous system.

Genes do not code for behaviour directly, but DNA variations that create differences in these physiological systems can affect behaviour. Genes do not blindly pump out their protein products. When the gene product is needed, many copies of its mRNA will be present, but otherwise very few copies of the mRNA are transcribed. You are changing the rates of transcription of genes for neurotransmitters by reading this sentence. Because mRNA exists for only a couple of minutes and then is no longer translated into protein, changes in the rate of transcription of mRNA are used to control the rate at which genes produce proteins.

In some cases, introns—parts of genes that are transcribed into RNA but spliced out before mRNA leaves the nucleus—regulate gene transcription. The sole function of many genes is to regulate the transcription of other genes rather than to code for proteins. Some gene regulation is short term and responsive to the environment. Other gene regulation leads to long-term changes in development. If a particular molecule binds with the regulatory sequence, it will free the gene for transcription.

Most gene regulation involves several mechanisms that act like a committee voting on increases or decreases in transcription. That is, several transcription factors act in concert to regulate the rate of specific mRNA transcription. Similar mechanisms also lead to long-term developmental changes. The key question about development is how differentiation occurs, how we start life as a single cell and end up with trillions of cells, all of which have the same DNA but many different functions.

Some basic aspects of development are programmed in genes. For example, we have 38 homeobox (*Hox*) genes similar to genes in most animals that act as master switches and control the timing of development of different parts of the body. However, for the most part, development is not hard wired in the genes. The point is that many genes regulate the transcription of other genes in response to both the internal and the external environments. It is now thought that much more DNA is involved in regulatory genes than in structural genes that code for proteins.

Although behavioural genetics has been concerned primarily with structural genes, its methods are just as appropriate for detecting genetic variation that arises from gene regulation. Behavioural genetics asks why people are different behaviourally—for example, why some are mentally ill or retarded. For this reason, it focuses on genetic and environmental differences among people that can account for these observed differences.

New DNA differences occur when mistakes, called *mutations*, are made in copying DNA. These mutations result in different alleles (called *polymorphisms*) such as the alleles responsible for the variations that Mendel found in pea plants, for Huntington's disease and PKU, and for complex behavioural traits such as schizophrenia and cognitive abilities. Mutations that occur in the creation of eggs and sperm will be transmitted faithfully unless natural selection intervenes. The effects that count in terms of natural selection are effects on survival and reproduction.

Because evolution has so finely tuned the genetic system, most new mutations in regions of DNA that are translated into amino acid sequences have deleterious effects. However, once in a great while a mutation will make the system function a bit better. In evolutionary terms, this means that individuals with the mutation are more likely to survive and reproduce. A single-base mutation can result in the insertion of a different amino acid into a protein. Such a mutation can alter the function of the protein.

For example, if the first DNA codon TAC is miscopied as TCC, the amino acid arginine will be substituted for methionine. This single amino acid substitution in the hundreds of amino acids that make up a protein might have no noticeable effect on the protein's functioning; then again, it might have a small effect, or it might have a major, even lethal, effect.

A mutation that leads to the loss of a single base is likely to be more damaging than a mutation causing a substitution, because the loss of a base shifts the *reading-frame* of the triplet code. Instead of the amino acid chain containing methionine (TAC) and leucine (AAC), the mutation would result in a chain containing serine (TCA) and tryptophan (ACC). Mutations are often not so simple. For example, a particular gene can have mutations at several locations. However, the CAG repeat responsible for Huntington's disease is transcribed into mRNA and translated into protein, which means that multiple repeats of an amino acid are inserted into the protein. CAG is the mRNA code, which means that the DNA code is GTC. GTC codes for the amino acid glutamine.

It is reasonable to expect that having a protein lumbered with many extra copies of glutamine would reduce the protein's normal activity; in other words, the lengthened protein would show loss of function. However, although Huntington's disease is a dominant disorder, the other allele should be operating normally, producing enough of the normal protein to avoid trouble. This possibility suggests that the Huntington's allele, which adds dozens of glutamines to the protein, might confer a new property that creates the problems of Huntington's disease.

DNA polymorphisms

Because millions of DNA base pairs are polymorphic, DNA polymorphisms can be used in linkage studies to track the chromosomal location of genes. As noted earlier, such DNA markers were first used to localize the gene for Huntington's disease at the tip of the short arm of chromosome 4. The first type of DNA marker has a long but descriptive name: *restriction fragment length polymorphism (RFLP).*

DNA extracted from blood or from cells scraped from the inside of the cheek is chopped up by a type of enzyme called a restriction enzyme. Restriction enzymes were found in bacteria, which produce them to defend against infecting viruses. There are hundreds of varieties of restriction enzymes, each of which cuts the DNA at a particular sequence, usually six base pairs in length, wherever that sequence is found. For example, one commonly used restriction enzyme, *EcoRI,* recognizes the sequence GAATTC and severs the DNA molecule between the G and A bases. This sequence occurs thousands of times throughout the genome.

If this DNA sequence differs at a particular locus for some individuals, their DNA will not be cut at that point. This failure to cut results in one long DNA fragment for these individuals; this long fragment can be identified when it is compared with fragments from individuals whose DNA is cut at that point into two shorter fragments. In other words, this is a polymorphism that can be detected when a restriction enzyme creates DNA fragments of different lengths from the DNA of two individuals. Although there are hundreds of restriction enzymes that detect different DNA sequences, in fact, they can detect only about one-fifth of all DNA polymorphisms.

Several other types of DNA markers have been developed to detect other DNA polymorphisms. The most widely used DNA marker, developed in 1987, is called a simple sequence repeat (SSR) marker, previously known as a microsatellite repeat marker. As mentioned in two, three, or four nucleotide base pairs are repeated dozens of times at as many as 50,000 loci throughout the genome. The number of repeats at each locus differs among individuals and is inherited in a Mendelian manner.

The human genome has been mapped with several thousands of very closely spaced SSRs to guide the search for genes. SSRs are especially useful because they can be detected by using machines called automated DNA sequencers, which can complete several thousand assays per day. Several other techniques are available to detect polymorphisms that cannot be identified by RFLP or SSR methods.

The ultimate way to detect all polymorphisms for a particular gene is to determine the DNA sequences for individuals. This task is becoming increasingly possible with automated DNA sequencers. Now that the appropriate techniques have been developed, DNA markers can be used to find genes associated with behavioural traits.

DNA replication

A dividing cell has to carry out extensive DNA replication, and must maintain a considerable degree of accuracy during this process. Even an error rate of 0.01% would cause a significant accumulation of alterations in the genes of a rapidly dividing organism such as a bacterium, quickly leading to the vital DNA sequences becoming meaningless. One of the final sentences of Watson and Crick's, paper on the double helix reads. It has not escaped our notice that the specific pairing we have postulated immediately suggests a possible copying mechanism for the genetic material! They were referring to the fact that complementary base pairing between the two polynucleotides of the double helix would enable each strand to act as a template for synthesis of its complement. Thus, the double helix structure possesses an obvious means of replicating itself. Nevertheless, molecular biologists were uncertain about the overall pattern of the process. Three different strategies for replication of the double helix seemed possible.

i) *Semi-conservative replication,* in which each daughter molecule contains one polynucleotide derived from the original molecule and one newly synthesized strand.

ii) *Conservative replication,* in which one daughter molecule contains both parent polynucleotides and the other daughter contains both newly synthesized strands.

iii) *Dispersive replication*, in which each strand of each daughter molecule is composed partly of the original polynucleotide and partly of newly synthesized polynucleotide.

Although the semi-conservative scheme seems most likely purely on intuitive grounds, and in fact was favoured by Watson and Crick, it was necessary to devise an experiment which would distinguish between these three modes of replication and confirm which scheme actually operates. Like many advances in biochemistry and molecular biology, the Meselson-Stahl experiment depended on the use of chemical isotopes. Nitrogen exists in several isotopic forms: as well as the 'normal' isotope, AN, which predominates in the environment and has an atomic weight of 14.008, there are a number of other isotopes that occur in much smaller amounts. These include ^{15}N, which because of its greater atomic weight is called 'heavy nitrogen'. Different isotopes can be distinguished from one another by a variety of means, so if a molecule containing an uncommon isotope is provided to a cell as a nutrient, the fate of the molecule (that is, the biochemical reactions that it undergoes in the cell) can be followed by identifying all the cellular metabolites that become labelled with the isotope.

Of relevance to the Meselson-Stahl experiment is the fact that *E. coli* cells provided with heavy nitrogen in the form of $^{15}NH_4$ Cl incorporate the labelled nitrogen into their DNA molecules. Heavy nitrogen is a stable isotope so does not emit radiation. However, DNA molecules containing heavy nitrogen can be distinguished from DNA molecules containing the normal isotope by density gradient centrifugation. This is because ^{14}N-DNA has a greater buoyant density than ^{14}N-DNA and therefore forms a band at a lower position in a CsCl density gradient. New polynucleotides synthesized after resuspension therefore contained only the normal isotope of nitrogen. The bacteria were then allowed to undergo one round of cell division, which takes roughly 20 minutes for *E. coli*, during which time each DNA molecule replicates just once. Some cells were then taken from the culture, their DNA purified and a sample analysed by density gradient centrifugation.

If we re-examine the three schemes for DNA replication we see that this result is compatible with both semi-conservative replication (each double helix comprises one ^{15}N-polynucleotide and one ^{14}N-polynucleotide) and dispersive replication (both strands are made up of a mixture of ^{14}N-polynucleotide and ^{14}N-polynucleotide). However, conservative replication can be ruled out at this stage, because this scheme would give two populations of DNA molecules, some entirely ^{14}N and some entirely ^{14}N: no hybrids would be seen.

To distinguish between semi-conservative and dispersive replication, Meselson and Stahl allowed the *E. coli* culture to undergo a second round of cell division. Again, cells were removed and their DNAs analysed in a density gradient. Two bands appeared: one representing the same hybrid molecules as before, the second corresponding to wholly ^{14}N-DNA. This result is entirely in agreement with the semi-conservative mode of replication, because according to this scheme there are now some granddaughter molecules that are made up entirely of ^{14}N-polynucleotides.

In contrast, the dispersive mode can be discounted because that method would still produce only hybrid molecules, and in fact would continue to do so for a very large number of cell generations. We can therefore conclude that DNA replication is semi-conservative. The Meselson-Stahl experiment is rightly considered a classic example of scientific technique. The problem to be resolved was carefully delineated, the available techniques were marshalled, and the experiment was designed so that the results would be unequivocal.

This masterpiece enabled the attention of biochemists and molecular biologists to move directly on to the second aspect of DNA replication, the question of which enzymes are involved and how the process actually occurs. When a DNA molecule is being replicated only a limited region is ever in a non-base-paired form. The breakage in base pairing starts at a distinct position, called the replication origin and gradually progresses along the molecule, possibly in both directions, with synthesis of the new polynucleotides occurring as the double helix unzips.

The important region, at which the base pairs of the parent molecule are broken and the new polynucleotides are synthesized, is called a replication fork. An enzyme able to synthesize a new DNA polynucleotide is called a DNA polymerase. The chemical reaction catalysed by a DNA polymerase is very similar to that of RNA polymerase except of course that the new polynucleotide that is assembled is built up of deoxyribonucleotide subunits rather than ribonucleotides.

In general terms the reactions are the same: the sequence of the new polynucleotide is dependent on the sequence of the template and is determined by complementary base pairing and, as with RNA synthesis, DNA polymerization can occur only in the 5'→3' direction. During DNA replication there is a continual need for breakage of the base pairing between the two strands of the parent DNA molecule. This is carried out by the DnaB helicase enzyme, which works in conjunction with single-strand binding proteins (SSBs). The latter attach to the single-stranded DNA that results from helicase action and prevent the two strands from immediately reannealing. The result is the replication fork that provides the templates on which DNA polymerase III can work. The main complication in DNA replication is that the two disassociated strands of the parent molecule cannot be treated in the same way. This is because DNA polymerase III can synthesize DNA only in the 5'→3' direction, which means that the template strands must be read in the 3'→-5' direction. For one strand of the parent, called the leading strand, this is no problem because the new polynucleotide can be synthesized continuously.

However, the second or lagging strand cannot be copied in a continuous fashion because this would necessitate 3'→5' DNA synthesis. Instead the lagging strand has to be replicated in sections: a portion of the parent helix is disassociated and a short stretch of the lagging strand replicated, a bit more of the helix is disassociated and another segment of the lagging strand replicated, and so on.

At first this process was just a hypothesis but the isolation of Okazaki fragments, short pieces of DNA between 100 and 1000 nucleotides in length, associated with DNA replication, confirmed that the suggestion is correct. Unlike RNA polymerase, DNA polymerase III cannot initiate DNA synthesis unless there is already a short double-stranded region to act as a primer. How can this occur during DNA replication? The answer appears to be that the very first few nucleotides attached to either the leading or the lagging strand are not deoxyribonucleotides but ribonucleotides that are put in place by an RNA polymerase enzyme. Once a few ribonucleotides (between six and 30) have been polymerized on the template the RNA polymerase moves aside, and polymerization, now of DNA, is continued by DNA polymerase III.

The RNA polymerase is called the primase and in *E. coli* is a single polypeptide with a molecular mass of about 60 000 daltons. It acts in conjunction with six or more additional proteins which make up a structure referred to as the primosome. On the lagging strand, DNA polymerase III can synthesize DNA only for a certain distance before it reaches the RNA primer at the 5' end of the next Okazaki fragment. Here DNA polymerase III stops and DNA polymerase I comes into action, continuing DNA synthesis and carrying out a further important function by removing the ribonuclecitides of the primer of the adjacent Okazaki fragment and replacing them with deoxyribonuclecitides.

When all the ribonucleotides are replaced, DNA polymerase I either stops or possibly carries on a short distance into the DNA region of the Okazaki fragment, continuing to replace nuclecitides before it dissociates from the new double helix. The final reaction needed to complete replication of the lagging strand is to join up adjacent Okazaki fragments, which are now separated by just a single gap between neighbouring nuclecitides. All that is needed is to synthesize a phosphodiester bond at this position, a reaction catalysed by another new enzyme, this one called DNA ligase. The system just described brings about replication of the leading and lagging strands of the parent molecule and produces two daughter double helices.

However, there is one outstanding problem to consider. The two polynucleotides of the parent helix are, of course, wound round one another; this means that progression of the replication fork along the parent molecule requires the double helix not just to be unzipped but also to be unwound. This is in fact a significant problem if you consider that the *E. coli* DNA molecule is 4000 kb, or 400 000 turns of the helix, in length and must be replicated in 20 minutes.

The implication is that the double helix is rotating at a rate of 20 000 rpm! This is so inconceivable that for many years molecular biologists sought solutions that avoided unwinding the helix. At one stage around things became so desperate that some scientists even suggested that the double helix was incorrect and that in fact the polynucleotides in a double-stranded DNA molecule are laid side by side and not wound round each other. Fortunately a group of enzymes that solve this topological problem was eventually discovered. The enzymes that unwind the double helix are called DNA topoisomerases and they fall into three classes, Type I-5', Type I-3' and Type II.

Each enzyme unwinds a DNA molecule without actually rotating the double helix; they achieve this feat by causing transient breakages in the polynucleotide backbone. Type I topoisomerases break just one of the polynucleotides and pass the other strand through the gap before reforming the backbone. Type II enzymes, which include the well-characterized DNA gyrase of *E. coli,* carry out the same sort of reaction but break both polynucleotides at adjacent positions while doing so. Despite this difference, all topoisomerases produce the same result: portions of the double helix just in advance of the replication fork are unwound allowing the fork to progress unhindered along the parent molecule.

DNA topoisomerases can also carry out the reverse reaction and introduce extra turns into a DNA molecule: this results in supercoiling as opposed to unwinding. DNA replication occurs in a similar manner in all organisms, and the scheme just described for events at the replication fork in *E. coli* also holds for eukaryotes. Mammalian cells have five different DNA polymerases, called α, β, γ, δ and ε, with DNA polymerase α and DNA polymerase δ being responsible for replication of nuclear DNA. Their precise roles are still not completely understood, but DNA polymerase δ probably replicates the leading strand, with the lagging strand being copied by DNA polymerase α.

The latter enzyme has a tightly associated primase activity, which is thought to be responsible for synthesizing the RNA primer for each Okazaki fragment. One puzzle is that neither DNA polymerase α nor DNA polymerase 8 have the 5'→3' exonuclease activities that are needed to remove the RNA primers from the lagging strand. Proofreading the lagging strand is also a problem as DNA polymerase a also lacks the 3→-5' exonuclease, and therefore cannot retrace its steps to replace an incorrect nucleotide.

Gene identification

During the past decade, however, quantitative genetics and molecular genetics have begun to come together to identify genes for complex, quantitative traits. Such a gene in multiple-gene systems is called a *quantitative trait locus (QTL).* In addition to producing indisputable evidence of genetic influence, the identification of specific genes will revolutionize behavioural genetics by providing measured genotypes for investigating genetic links between behaviours, for tracking the developmental course of genetic effects, and for identifying interactions and correlations between genotype and environment. Moreover, once a gene is identified, it is possible to determine what protein the gene produces and to investigate how the gene's product affects behaviour. In addition to studying naturally occurring genetic variation, geneticists have used chemicals to create mutations that enable them to dissect the effects of genes. This mutational strategy has been applied to the study of behaviour.

During the past 30 years, hundreds of behavioural mutants have been selected in organisms as diverse as single-celled organisms, worms, fruit flies, and mice. This work illustrates the point that most normal behaviour is influenced by many genes. Although any one of many single-gene mutations can seriously disrupt behaviour, normal development is orchestrated by many genes working together. An analogy is an automobile, which requires thousands of parts for its normal functioning. If any one part breaks down, the automobile may not run properly. In the same way, single genes can drastically affect behaviour that is normally influenced by many genes.

Although the behaviour of bacteria is by no means attention grabbing, they do behave. They move toward or away from many kinds of chemicals by rotating their propellerlike flagella. Since the first behavioural mutant in bacteria was isolated in the dozens of mutants that have been created emphasize the genetic complexity of an apparently simple behaviour in a simple organism. For example, many genes are involved in rotating the flagella and controlling the duration of the rotation. Like bacteria, paramecia are one-celled organisms, but they are larger and their movements are more obvious. Propelled by cilia, paramecia avoid certain chemicals and heat by backing up and then swimming forward in a new direction. Hundreds of induced behavioural mutants have been isolated. For example, at least 20 genes are involved in the avoidance behaviour of paramecia. Some mutants that cannot swim backward are called *pawn* mutants (after the chess piece that can only move forward). Mutations at any of three loci may result in this behaviour. Other mutants include *paranoiac* (prolonged backward movement), which involves mutations at five loci, and *sluggish* (involving very slow movement), which is due to a single locus. The nematode (roundworm) *Caenorhabditis elegans* has 959 cells, of which 302 are nerve cells. Its behaviour is more complex than that of singlecelled organisms like bacteria and paramecia, and many behavioural mutants have been identified. For example, several mutations that affect locomotion and foraging behaviour have recently been identified. *C. elegans is* especially important for behavioural genetic analysis because the developmental fate of each of its cells, the wiring diagram of its 302 nerve cells, and most of its several thousand genes are known. The complete description of its DNA sequence of 100 million base pairs is expected soon.

The fruit fly *Drosophila* is the star organism in terms of behavioural mutants, with hundreds known. The earliest behavioural research involved response to light (phototaxis) and to gravity (geotaxis). Normal *Drosophila* move toward light (positive phototaxis) and away from gravity (negative geotaxis). Many mutants that were either negatively phototaxic or positively geotaxic were created. The hundreds of other behavioural mutants included *sluggish* (generally slow), *hyperkinetic* (generally fast), *easily shocked* (jarring produces a seizure), and *paralysed* (collapses when the temperature goes above 28°C). A *drop dead* mutant walks and flies normally for a couple of days and then suddenly falls on its back and dies.

More complex behaviours have also been studied, especially courtship and learning. Behavioural mutants for various aspects of courtship and copulation have been found. One male mutant, called *fruitless,* courts males as well as females and does not copulate. Another male mutant cannot disengage from the female after copulation and is given the dubious tide *stuck.* The first learning behaviour mutant was called *dunce* and could not learn to avoid an odour associated with shock even though it had normal sensory and motor behaviour. Other learning mutants have been found as well as a memory mutant called *amnesiac. Drosophila* also offer the possibility of creating genetic mosaics in which individuals have different genes in various cells of the body. This method has been applied to many behavioural mutants to analyse which body parts must be altered to allow the mutant behaviour to occur.

In an early study, mosaics were created with one eye with a normal gene and positive phototaxis and one eye with a mutant gene and nonphototaxis. When a light is placed at the top of the tube, the mosaic flies attempt to equalize the light coming into the two eyes and thus keep their defective eye toward the light. As a result, the mosaic flies climb up the tube in a spiral, keeping their mutant eye closer to the light. The most interesting mosaic mutant studies involved sexual behaviour and the X chromosome.

Drosophila were made mosaic for the X chromosome so that some body parts have two X chromosomes and are female and other body parts have only one X chromosome and are male. As long as a small region toward the back of the brain is male, courtship behaviour is male. Of course, sex is not all in the head. Different parts of the nervous system are involved in aspects of courtship behaviour such as tapping, "singing," and licking. Successful copulation also requires a male thorax (containing the fly's version of a spinal cord between the head and abdomen) and, of course, male genitals. Most amazing is the ability to add bits to a gene that act as a switch, turning the gene on or off (i.e., creating inducibility), for example, when a particular hormone is added or even when the temperature changes. For mice, most gene targeting research is cruder.

A gene is "knocked out" entirely by deleting some of its key DNA sequences. Once mice homozygous for the knock-out gene are bred, the effect of the knock-out gene on behaviour can be investigated. One of the first experiments of this kind for behaviour produced mice with a knock-out version of a gene (α-CaMKII) for a protein enriched in the hippocampus and thought to be involved in long-term memory. Mutant mice homozygous for the knock-out gene learned a spatial task significantly more poorly than control mice, although otherwise their behaviour seemed normal. Knock-out alleles of several other genes have been shown to affect learning and memory. Another example relevant to behaviour involved aggression. Increased aggression was shown by male mice with a knock-out version of a gene that codes for a receptor for an important neurotransmitter, serotonin. Increased aggression was also found for another knock-out of a gene for an enzyme (neuronal nitric oxide synthase) that normally plays a basic role in neuro-transmission.

Hundreds of knock-out studies are in progress and many associations with behaviour are likely to emerge. The significance of these associations between single-gene mutations and behaviour is that knocking out a single gene can drastically affect behaviour that is normally influenced by many genes.

Quantitative trait locus linkage

Complex quantitative traits are likely to be influenced by multiple genes as well as by multiple environmental factors. Finding such QTLs has only become possible in recent years. Animal models have been particularly useful. For example, researchers have identified 23 QTLs that account for most of the genetic variation in diabetes in the mouse. The QTLs have various effect sizes. More than half of the genetic variation is due to interactive effects between QTLs. Work on rat models of hypertension shows a similar complex picture. Allelic association between candidate genes and particular phenotypes is a simple and powerful approach to QTL identification. However, it is usually difficult to know which of the thousands of genes to investigate as possible candidate genes affecting behaviour.

Finding allelic association requires that a DNA marker must itself be the gene that causes the association or that the marker must be so close to the gene that the marker and the gene are not separated by recombination. In contrast to allelic association, linkage can be used to scan die entire genome with just a few hundred markers. Linkage is an exception to Mendel's second law of independent assortment. If a DNA marker is on the same chromosome and within the general chromosomal region of a gene that affects a behavioural disorder, the marker and disorder will not assort independently, thus indicating linkage.

However, behavioural dimensions and disorders are likely to be influenced by many genes. If many genes contribute to behaviour, behavioural traits will be distributed quantitatively. The goal is to find some of the many genes (QTLs) that affect these quantitative traits. Allelic association is able to detect small QTL effects, but only recently have linkage approaches been able to identify QTLs. For example, QTL linkage strategies have been applied to open-field activity in mice. One QTL strategy involves F_2 mice derived from a cross between the high and low lines selected for open-field activity and subsequently inbred by using brother-sister matings for over 30 generations. Each F_2 mouse has a unique combination of alleles from the parental strains.

The most active and the least active F_2 mice were examined for 84 DNA markers spread throughout the mouse chromosomes in order to identify chromosomal regions that are associated with open-field activity. The analysis simply asks whether the frequencies of various alleles differ between the most active and least active groups. Although this analysis is like allelic association, F_2 animals have an average of only one recombination per chromosome, so a marker can detect linkage with a gene millions of base pairs away on a chromosome. A QTL on chromosome 15 is related primarily to open-field activity and not to other measures of fearfulness, an observation suggesting the possibility of a gene specific to open-field activity.

The QTL regions on chromosomes 1 and 12, on the other hand, are related to other measures of fearfulness, an association suggesting that the same QTL affects these diverse measures of fearfulness. The exception is exploration in an enclosed arm of a maze, which was included-in the study as a control, because other research suggests that this measure is not genetically correlated with measures of fearfulness. Another method used to identify QTLs for behaviour involves special inbred strains called *recombinant inbred (RI) strains.* The chromosomes of RI strains have been examined and mapped for more than a thousand markers, thus enabling investigators to determine whether any of these markers are linked to behaviour.

For example, an RI QTL analysis of open-field activity suggests a QTL on chromosome 6 and possible QTLs on other chromosomes. Various QTL approaches are being applied to many domains of behaviour, especially in the area of psychopharmacogenetics. For statistical reasons, only the largest QTL effects can be detected by these methods. Although many QTLs are likely to have much smaller effect sizes, it makes sense to begin the quest for QTLs with the biggest QTL effects. Especially exciting is the possibility of using QTLs found in mice as candidate QTLs for human research. Specific genes that are associated with behaviour in mice can be used in human studies because nearly all mouse genes are similar to human genes.

Moreover, chromosomal regions linked to behaviour in mice can be used as candidate regions in human studies because parts of mouse chromosomes have the same genes in the same order as parts of human chromosomes, a relationship called *synteny homology*. In our species, experimental crosses such as those producing recombinant inbred strains cannot be used to identify genes for behaviour. However, as in research on animal behaviour, allelic association studies can be used to examine the effects of candidate genes on behaviour. For example, a particular allele of a gene (apolipoprotein E on chromosome 19) involved in cholesterol transport is associated with late-onset Alzheimer's disease. This allele quadruples the risk for Alzheimer's disease.

Although the association remains controversial, a gene for one of the neuronal receptor sites for dopamine (dopamine receptor D_2 on chromosome 11) has been associated with alcoholism and other drug abuse. Allelic associations are beginning to be reported for psychopathology, personality, and cognitive abilities. Allelic association is especially useful for detecting small QTL effects. As mentioned earlier, in relation to mouse behaviour, a systematic search of the genome for genes for behaviour requires linkage rather than allelic association.

For single-gene disorders, linkage can be identified by using large family pedigrees in which cotransmission of a DNA marker allele and a disorder can be traced. Because recombination occurs an average of once per chromosome in the formation of gametes passed from parent to offspring, a marker allele within millions of base pairs and a gene for a disorder on the same chromosome will be inherited together within a family. Linkage research has been boosted by the construction of systematic maps of landmarks, called sequence-tagged sites (STSs), that cover the entire human genome. The human genome map now includes more than 15,000 STSs, at an average interval of less than 200,000 base pairs. In this family, the gene for Huntington's disease is linked to the allele labelled C.

All but one person with Huntington's disease has inherited a chromosome that happens to have the C allele in this family. This marker is not the Huntington's gene itself, because a recombination was found between the marker allele and Huntington's disease for one individual; the leftmost woman shown in generation IV had Huntington's disease but did not inherit the C allele for the marker. That is, this woman received that part of her affected father's chromosome carrying the gene for Huntington's disease, which is normally linked in this family with the C allele but in this woman is recombined with the A allele from the father's other chromosome. The farther the marker is from the disease gene, the more recombinations will be found within a family. Markers even closer to the Huntington's gene were later found.

A genetic defect was identified as the CAG repeat sequence associated with most cases of Huntington's disease. Precisely how the CAG repeat causes Huntington's disease is not yet known. A similar approach was used to locate the genes responsible for-other single-gene disorders such as PKU on chromosome 12 and fragile X mental retardation on the X chromosome. Although linkage analysis of large pedigrees has been very effective for locating genes for single-gene disorders, it is less powerful when several genes are involved.

Even for complex disorders like schizophrenia, for which many genes may be involved in the population, the traditional large-pedigree design might be successful in identifying some families in which a particular gene has a large effect. For example, schizophrenia researchers are currently focusing on the end of the short arm of chromosome 6 (6*p*24-22), because a pedigree study reported linkage there for about a quarter of families with multiple schizophrenic members. Another linkage approach is less restrictive than the traditional large-pedigree design. It examines *allele sharing* for pairs of affected relatives, usually siblings, in many different families.

For a particular DNA marker, a sibling pair can share 0, 1, or 2 of their parents' alleles. Half of sibling pairs are expected to share a particular marker allele. If both members of a sibling pair are affected, they should show greater than 50 percent allele sharing for markers in a region that harbours a gene that contributes to risk for schizophrenia. In other words, linkage is indicated in an affected sib-pair linkage study when markers in a region show allele sharing that exceeds the expected 50 percent. The affected sib-pair linkage design also was used to identify and replicate a linkage on the X chromosome for homosexuality inherited through the mother.

Allele sharing for markers in the *q*28 region of the X chromosome was 67 percent for homosexual brothers. In contrast, allele sharing for the heterosexual brothers of these homosexual men was only 22 percent, significantly less than the expected 50 percent allele sharing, again suggesting linkage between sexual preference and the X chromosome. Linkage based on allele sharing can also be investigated for quantitative traits by correlating allele sharing for DNA markers with sibling differences on a quantitative trait. That is, a marker linked to a quantitative trait will show greater than expected allele sharing for siblings who are more similar for the trait.

The sib-pair QTL linkage design was first used to identify and replicate a linkage for reading disability on chromosome 6. It is clear that the field of psychology is at the dawn of a new era in which behavioural genetic research is moving beyond the demonstration of the importance of heredity to the identification of specific genes.

In clinics and research laboratories, psychologists of the future will routinely collect cells from the inside of the cheek and send them to a laboratory to extract DNA and to genotype specific genes associated with particular psychological traits. This is already happening for apolipoprotein E and late-onset dementia and for fragile X mental retardation in males. As is the case with most important advances, identifying genes for behaviour will raise new ethical issues that are already affecting genetic counselling. Identifying genes for single-gene disorders has led to concerns about employment and insurance discrimination. The benefits of identifying genes for understanding the etiology of behavioural disorders and dimensions seem likely to far outweigh the potential abuses.

Biological information and gene expression

Biological information carried by a gene is contained in its nucleotide sequence. This information is in essence a set of instructions for the synthesis of an RNA molecule that may subsequently direct the synthesis of an enzyme or other protein molecule, or may itself be functional in the cell. This process is called *gene expression*.

At this stage it is sufficient to appreciate that the nucleotide sequence of the gene is the crucial factor as far as the biological information is concerned. Genes are discrete segments of DNA molecules, separated from one another by *intergenic DNA*. Genes are arranged in different ways in different types of organism. In viruses, for example, genes are closely packed with very little intergenic DNA between them (indeed, some genes may overlap). On the other hand, in higher organisms the genes are more spread out and are often separated by very long intergenic regions.

In humans the genes make up less than 30%, of the total DNA in the cell. The biological information of a gene is carried by just one of the two polynucleotides of the double helix. This polynucleotide is called the *template strand* because it acts as a template for synthesis of a complementary RNA molecule during the first stage of gene expression. The biological information carried by the template strand is read in the 3'→5' direction. Two genes on the same DNA molecule do not necessarily have the same polynucleotide as their template strand.

The information-carrying capability of genes is almost unlimited. Remember that a gene 150 bp in length, which is in fact quite short, could have any one of 4^{150} different nucleotide sequences. In practice, not all, of these 'genes' would be meaningful in biological terms because there are certain rules that place limitations on the number of sequences that make sense. Nevertheless, a large proportion of the 4^{150} sequences would contain biological information. There is no difficulty in rationalizing the known nature of the gene with the vastness of biological variability. The majority of genes are spaced out more or less randomly along the length of a DNA molecule. In some cases, however, they are grouped into distinct clusters. Sometimes the individual genes in a clusters. As is the convention when cluster are unrelated and there is no apparent reason or advantage to having them organized in this way.

More frequently clusters are made up of genes that contain related units of biological information. Two examples are the *operon* and the *multigene family*. Operons are fairly common features of the organization of genes in bacteria. An operon is a cluster of genes coding for a series of enzymes that work in concert in an integrated biochemical pathway. The first operon to be discovered was the lactose or lac operon of *E. coli,* a cluster of three genes each coding for one of the three enzymes involved in conversion of the disaccharide lactose into its monosaccharide units glucose and galactose.

These enzymes are not required all the time, only when lactose is available to the bacterium. But when needed they are needed together: the three units of biological information carried by the lac genes must be read at the same time. A sophisticated system enabling the lac genes to be expressed together and only when needed has evolved round the fact that the genes are clustered.

Operons have no direct counterparts in organisms other than bacteria. In contrast, multigene families are found in many organisms. A multigene family is also a cluster of related genes, but in multigene families the individual genes have identical or similar nucleotide sequences, and therefore contain identical or similar pieces of biological information. There are two types of multigene family:

i) *Simple multigene families,* in which all of the genes are exactly or virtually the same. For example, most higher organisms have multiple copies of the gene for the 5S rRNA, probably because large quantities of 5S rRNA molecules must be synthesized at certain times, creating a demand that just one or a few genes would be unable to satisfy. Rather than being spread randomly throughout the chromosomes, the 5S rRNA genes are clustered into a multigene family. In humans, for example, there is a family of about two thousand 5S rRNA genes on chromosome 1.

ii) *Complex multigene families* are made up of similar but non-identical genes. A good example is provided by the gene family for the globin polypeptides of vertebrates. In humans four globin polypeptides, 2α and 2β, combine with a haem cofactor to make a single molecule of haemoglobin, the protein responsible for transporting oxygen in the bloodstream. Each type of globin polypeptide, α and β, exists as a family of related molecules differing from one another at just a few amino acid positions within the polypeptide chain. In humans there are in fact two α-like globins, α and ξ and five β-like globins, β itself, δ, $^{A}\gamma$, $^{G}\gamma$ and ε. The types of globin found in the bloodstream depend on the stage of development of the organism, with ξ and ε, for instance, found only in the early embryo.

The human globin genes are clustered into an a multigene family on chromosome 16 and a b multigene family on chromosome 11. Subsequent analysis has allowed the precise positions of the individual genes in the clusters to be mapped.

We now have information on a whole range of multigene families in a variety of organisms, including some families in which the individual genes are not in fact clustered but scattered at different positions on a single or on more than one chromosome. This indicates that clustering of members of a multigene family is not an aid to coordinated expression of the genes but probably is merely a result of the evolutionary processes that gave rise to the family. We can speculate that the members of each family arose from an ancestral gene by a series of duplications, followed by events that have slightly altered the sequences of the resulting genes. Large-scale rearrange-ments within and between DNA molecules could subsequently break the cluster and scatter the members of the family. Much of what we know about gene structure has been gained by the application of recombinant DNA techniques.

The main types of experimental manipulation and the sort of information that can be obtained:

i) DNA molecules can be Cleaved into defined segments. DNA molecules can be cut into fragments by enzymes called *restriction endonucleases*, which cleave DNA only at certain nucleotide sequences. A single gene, or group of genes, can therefore be obtained as a small fragment of DNA.

ii) DNA fragments can be purified. DNA fragments can be separated by gel electrophoresis and individual ones recovered in pure form from the gel.

iii) Cloning provides Large quantities of a chosen gene. DNA fragments can be inserted into a *cloning vector* and large amounts obtained by gene cloning. Usually the entire DNA content of an organism, rather than a purified gene, is used as the starting material for a cloning experiment. If this is the case then many different clones, each containing a different fragment of DNA, are obtained. This *clone library* can then be screened to identify the clone containing the gene of interest.

iv) PCR can also provide large quantities of a chosen gene. As an alternative to cloning, a defined segment of DNA can be amplified into multiple, identical copies by the *polymerase chain reaction* (PCR). As with cloning, the experiment can be designed so that a selected gene is obtained, even if the starting material is the entire DNA content of a cell.

v) The nucleotide sequence of a gene can be determined. Using modern techniques a sequence of several kb (an average-sized gene) can be worked out in a few weeks, and much longer sequences (several Mb) can also be obtained.

Globin multigene families illustrate a second feature of genes and gene organization: the *pseudogene*. As well as the genes coding for the α-like and β-like globins, the human gene clusters also include additional genes, labelled $\psi\xi_1$, $\psi\alpha_1$, $\psi\alpha_2$ and $\psi\beta_2$. These 'genes' are very similar to the other members of their family but we can tell from their nucleotide sequences that the biological information they contain has become scrambled and is no longer readable.

Apparently minor changes in a gene's nucleotide sequence can completely destroy the meaning of the biological information, leaving a pseudogene that can no longer perform its function. It appears that once the biological information is lost a pseudogene undergoes sequence alterations until it is no longer recognizable as a gene relic. More and more pseudogenes are being found as the DNA molecules of lower and higher organisms are studied in increasing detail with the sophisticated techniques now available to the molecular biologist.

Discontinuous genes

Discontinuous genes (also called split or mosaic genes) are now known to be common in higher organisms and many types of virus, but until recently were thought to be absent in bacteria. This is probably the case with eubacteria (the vast majority of known species) but the latest research has revealed introns in certain genes of the ancient organisms called archaebacteria. In higher organisms a gene may contain no introns or may have as many as a hundred or more. Often the introns are much longer than the exons, a point that is illustrated by the human gene for the cystic fibrosis transmembrane regulator. This gene, which can cause cystic fibrosis when it does not function correctly, is 250 kb from beginning to end and is split into 24 exons and 23 introns. The average length of the exons is 227 bp, so all the exons added together make up only 2.4% of the gene. These exons are scattered throughout the entire length of the gene, separated by introns that range in size from 2 to 35 kb.

Why some genes should be split into exons and introns has been the subject of intense speculation? An interesting, although not necessarily correct, theory is that each exon of a discontinuous gene contains a different subcomponent of the biological information carried by the gene as a whole. Individual subcomponents are not sufficient on their own to code for a complete protein, but the information they contain is meaningful in that it specifies a recognizable portion of the protein's function, for example enabling the protein to bind a particular substrate or attach to a specific site in the cell. The hypothesis is that, during evolution, exons from different discontinuous genes can be 'shuffled', creating new combinations of biological information.

Exon shuffling would be more likely to produce new functional proteins than would an entirely random rearrangement of existing genes. The discovery of exons has led to many speculations concerning the evolution of genes before the origin of living cells. One theory is that the first 'genes' to form in the prebiotic soup were not continuous sequences of nucleotides, but instead were made up of a number of short mini-genes, each containing a small fragment of biological information. Initially these mini-genes would have been separated by vast regions of meaningless nucleotide sequence, or possibly were even on different molecules, but eventually they became associated together to form a discontinuous gene, with the mini-genes becoming the exons.

These earliest genetic molecules may have been DNA, but it now seems more likely that they were RNA, with DNA copies being formed at a subsequent stage in evolution. If genes were formed in this way then bacteria must have originally possessed introns, losing them at a later stage, possibly to streamline their DNA molecules, enabling more rapid DNA replication and a shorter generation time between one cell division and the next. For all genes the end-point of gene expression is synthesis of either an RNA molecule or a polypeptide.

How can this simple process bring about the utilization of the biological information contained in genes and thereby enable construction of a living organism? The answer lies in the functional flexibility of protein molecules. This is because polypeptides of different amino acid sequences can have quite different chemical properties and can therefore play quite different roles in the cell. The development and functioning of a living organism can be looked on as nothing more than the coordinated activity of a vast range of different protein molecules. The biological information contained in the genes, the blueprint for life itself, is simply the instructions for synthesizing these proteins at the correct time and in the correct place.

Entire complement of genes in a single cell represents a staggering amount of biological information. Some of this information is needed by the cell at all times: for instance, most cells continually synthesize ribosomes and so have a continuous requirement for transcription of the rRNA and ribosomal protein genes. Similarly, genes coding for enzymes such as RNA polymerase or those involved in the basic metabolic pathways are active in virtually all cells all of the time.

Genes that are active all the time are called housekeeping genes, reflecting the role in the cell of the biological information they carry. On the other hand, many genes have more specialized roles and their biological information is needed by the cell only under certain circumstances. All organisms are therefore able to regulate expression of their genes, so that those genes whose RNA or protein products are not needed at a particular time are switched off. This is a straightforward concept but it has extensive ramifications.

To illustrate the point we will briefly consider the basic features of gene regulation in bacteria and in eukaryotes. Gene regulation enables a bacterium to respond to changes in its environment. An example is provided by those bacterial genes involved in utilization of different sugar compounds as sources of carbon and energy.

E. coli possesses a number of genes coding for a variety of enzymes that enable the cell to take up and metabolize any of several different sugars. Exactly which enzymes are required by an individual bacterium at any one time depends on which sugars are present in the environment. Sometimes there is a plentiful supply of glucose, at other times the glucose levels are low-and the bacterium must utilize a different sugar, lactose for instance. *E. coli* could continuously express all its sugar utilizing genes, and so have molecules of each enzyme available all the time, but this would waste energy, something all organisms try to avoid.

Why synthesize enzymes for lactose utilization if there is no lactose available? Instead, a bacterium expresses only those genes coding for the enzymes needed to metabolize the sugars immediately available; the genes for all the other enzymes are inactive, switched off as their gene products are not required at the present time. If the environment changes, and one sugar is replaced by another, the bacterium can quickly switch on expression of the genes whose products are now needed, and switch off the ones that have become redundant.

So by regulating expression of their genes, *E. coli* and other bacteria are able to respond quickly to changes in the environment without wasting energy by maintaining in an active state genes whose products are of no immediate use. At the most basic level, gene regulation achieves the same thing in both baeteria and eukaryotes: it enables a cell to change its biochemical capabilities.

However, with eukaryotes we see a greater degree of sophistication with respect to both the signals that influence gene expression and the impact that gene regulation has on the organism. Eukaryotic cells respond to a *greater range of regulatory stim*uli. As-with *E. coli,* eukaryotic cells respond to changes in the environment by altering their gene expression patterns. For example, yeast regulates its genes for sugar utilization enzymes in a manner analogous to *E. coli,* and plant cells switch on genes for photosynthetic proteins in response to light.

In a multicellular organism, individual cells, or groups of cells, also respond to stimuli that originate from within the organism. Hormones, growth factors and other regulatory molecules ire produced by one type of cell and cause changes in gene expression in other cells. In this way the activities of groups of cells can be coordinated in a manner that is beneficial to the organism as a whole. Some eukaryotic genes are developmentally regulated. Most eukaryotes switch genes on and off at different stages of development.

Gene regulation in multicellular organisms results in cell specialization. In humans there are about 250 cell types, each with a different morphology, biochemistry and role to play in the organism. The distinctions between cell types are the result of differences in gene expression patterns. A liver cell is very different from a muscle cell because it expresses a different set of genes.

These gene expression patterns are permanent—a liver cell can never become a muscle cell—and are laid down very early in development. Inevitably, this greater degree of sophistication means that the gene regulation processes in eukaryotes are more complex than those in bacteria. There are, however, important similarities and we can understand a great deal about eukaryotic gene regulation by examining the control of gene expression in bacteria.

Control of gene expression is, in essence, control over the amount of gene product present in the cell. This amount is a balance between two factors: the rate of synthesis and the degradation rate. The result of this balance is a different steady state concentration for each gene product in the cell: if either the synthesis rate or the degradation rate changes then the steady state concentration also changes.

Genetic code

At the most fundamental level the structures of polypeptides and nucleic acids are the same: both are polymers made up of a series of linked monomers. In a polypeptide the monomers are called amino acids and the polymeric chain is generally less than 2000 units in length, much shorter than most naturally occurring nucleic acid molecules. Twenty different amino acids are found in protein molecules. The R groups vary considerably in chemical complexity: for glycine the R group is simply a hydrogen atom, whereas for tyrosine, phenylalanine and tryptophan the R groups are complex aromatic side-chains. The majority of R groups are uncharged, though two amino acids have negatively charged R groups (aspartic acid and glutamic acid) and three have positively charged R groups (lysine, arginine and histidine). Some R groups are *polar* (e.g. glycine, serine and threonine), others are *non-polar* (e.g. alanine, valine and leucine).

These differences mean that although all amino acids are closely related, each has its own specific chemical properties. The polymeric structure of a polypeptide is built by linking a series of amino acids by peptide bonds, formed by condensation between the carboxyl group of one amino acid and the amino group of a second amino acid. The structure of a tripeptide, comprising three amino acids. As with polynucleotides, the two ends of a polypeptide are chemically distinct: one has a free amino grout) and is called the amino-, NH_2-, or N-terminus; the other has a free carboxyl group and is called the carboxyl-, COOH-, or C-terminus.

In protein molecules, four levels of structure are recognized. *The primary structure,* which is the amino acid sequence itself. *The secondary structure,* which is the regular or repeating configuration taken up by the amino acid chain. The two most important secondary structures are the *α-helix* and *β-sheet*. Both are stabilized primarily by hydrogen bonding between the carboxyl and amino groups of different amino acids. Usually different regions of a polypeptide take up different secondary structures, so that the whole is made up of a number of a-helices and b-sheets, together with less organized regions. *The tertiary structure,* which is the three-dimensional conformation formed by folding together the secondary structural components of the polypeptide. The tertiary structure is held together by a variety of interactions between different amino acids, including hydrogen bonds, covalent linkages called *disulphydryl bridges* which can form between two cysteine residues, and the natural tendency of the polypeptide chain to fold up so that non-polar R groups are shielded from water. *The quaternary structure,* which refers to the way in which two or more polypeptides are orientated together to form a multi subunit structure. The quaternary structure may involve several molecules of the same polypeptide or may comprise different polypeptides, as with RNA polymerase. Each of the higher levels of protein structure—secondary, tertiary and quaternary—is specified by the primary structure, the amino acid sequence itself. This is most clearly understood at the secondary level, where it is recognized that certain amino acids, because of the chemical and physical properties of their R groups, stimulate the formation of an α-helix, whereas others promote formation of a β-sheet.

Conversely, certain amino acids more frequently occur outside regular structures and may act to determine the end-point of a helix or sheet. These factors are now so well understood that rules to predict the secondary structures taken up by amino acid sequences have been developed.Several kinds of experiment can be carried out in order to study the structure a protein. The amino acid composition can be determined by breaking the protein into its constituent amino acids, usually by placing a sample in hot, concentrated acid for several hours. The free amino acids can then be separated and quantified by chromatography. The amino acid sequence can be worked out by any of several related procedures. These involve a stepwise degradation process, with a single amino acid at a time being cleaved from the amino-terminus of the polypeptide. This amino acid is identified by its chromatographic properties and the cycle then repeated. The procedure continues until the entire amino acid sequence has been determined. Automatic polypeptide sequencers have been available.

Through X-ray diffraction analysis the tertiary structure can be investigated. Recently other techniques have also been developed, including nuclear magnetic resonance (NMR) spectroscopy, which allows the environment of different chemical nuclei to be determined. This technique was first developed for structural studies of small molecules, but has gradually been refined so that it can now provide information on molecules as large as proteins. Although less well characterized, it is nonetheless clear that the tertiary and quaternary structures of a protein also depend on the amino acid sequence. The interactions between individual amino acids at these levels are so complex that predictive rules, although attempted, are still unreliable.

However, it has been established for some years that if a protein is denatured, for instance by mild heat treatment or adding a chemical denaturant such as urea, so that it loses its higher levels of structure and takes up a non-organized conformation, it still retains the innate ability upon renaturation (by cooling down again, for example) to reform spontaneously the correct tertiary structure.

Once the tertiary structure has formed, subunit assembly into a multimeric protein again occurs spontaneously. This shows that the instructions for the tertiary and quaternary structures must reside in the amino acid sequence. From these considerations it is only a small step to appreciate that the function of a protein is also determined by its amino acid sequence. As an illustration we will consider those proteins that must attach themselves to a DNA molecule in order to perform their function in the cell. These DNA-binding proteins form a large and diverse group that includes, for instance, RNA polymerase and a number of important regulatory proteins which modulate, and at times block, transcription of individual genes. An example is provided by Cro, a regulatory protein that controls expression of a number of genes of the bacteriophage called λ. Cro is a dimer of identical polypeptides, each made up of three α-helices and a region of β-sheet.

In the active form of the protein, the two α_3 helices are exactly 34 Å apart and therefore fit into two adjacent sections of the major groove of a DNA molecule. If the α_3 helices were absent, or orientated incorrectly, the DNA-binding ability of the protein would be lost. DNA binding depends therefore on the protein assuming the correct secondary, tertiary and quaternary structures, which, in turn, depends on the amino acid sequence. This is a crucial fact in molecular genetics: the function of a protein depends on its amino acid sequence which, in turn, is specified by the sequence of nucleotides in the mRNA, which is itself a copy of the gene.

The biological information carried by the gene codes, in essence, for a protein function. Having grasped this important point we can now return to the question of the genetic code. The most enlightened molecular biologists, centred as usual around Crick, strove to prevent the ideas about the code becoming too complex.

Gradually the simplest system that was compatible with the established facts emerged as a working hypothesis. This system was based on two assumptions, both of which were subsequently confirmed by experimental analysis. It was more or less assumed right from that genes are colinear with the proteins they code for, meaning that the order of nucleotides in the gene correlates directly with the order of amino acids in the corresponding polypeptide. This is clearly the most straightforward way in which genes could code for proteins but was not in fact proved experimentally. The type of experiment needed was fairly obvious: alter the nucleotide sequence of a gene at a specific point and determine whether the resulting change in the amino acid sequence of the corresponding polypeptide occurs at the same relative position or elsewhere. The problem was that although at that time alterations could be introduced into genes without too much trouble, determining exactly where in the gene an alteration had occurred was much more difficult.

Eventually Charles Yanofsky and his colleagues managed to carry out the necessary analysis, with the *E. coli* gene coding for the enzyme called tryptophan synthase. The result was a clear demonstration of colinearity between gene and protein. Very shortly afterwards, Sydney Brenner's group completed a similar analysis of a gene and its protein from T4 phage, confirming Yanofsky's result. Later it was shown that the amino-terminus of the protein corresponds to the 5' end of the gene.

Codons cannot be just single nucleotides (A, T, G or C) because that would allow only four different codewords when 20 are required, one for each of the 20 amino acids found in proteins. Similarly, a doublet code (codons such as AT, TA, TT, GC, etc.) can be ruled out as this would contain only $4^2 = 16$ different codons. However, the next stage up, a triplet code (codons AAA, AAT, TAT, GCA, etc.) would be feasible as this would yield $4^3 = 64$ codewords, which would be more than enough.

A group of chemicals called the *acridine dyes*, of which *proflavin* is an example, cause single base pair deletions or additions in double-stranded DNA molecules. It was also known that certain proteins contain segments where the amino acid sequence can be changed without altering the function of the protein, although other regions of the same protein will not tolerate such changes. What if a series of insertions and/or deletions were introduced into the region of a gene coding for a tolerant segment of a protein?

If the code is triplet then a single insertion or deletion would give rise to a non- functional protein, because all the codewords downstream of the mutation would be altered, including those in the non-tolerant segment following the tolerant region. Two insertions or deletions would have the same effect. But three insertions or deletions would restore the correct *reading frame* in the non-tolerant region and would be predicted to have no effect on protein function.

An elegant experiment of this kind was eventually carried out successfully by Crick, Brenner and others, using the *rIIb* gene from T4 phage as the target, and determining the effect of proflavin-induced alterations by observing whether or not the treated phage were able to infect *E. coli*. Gradually, as the progressed, technical advances were made that would eventually allow the question of which triplet codon specifies which amino acid to be answered.

If translation could be directed by an RNA molecule of known nucleotide sequence, then it would be possible to assign individual codons by looking at the amino acid sequence of the protein that is synthesized. To do this it would be necessary to make artificial RNA molecules of known or predictable sequence. This became possible with the discovery by Severo Ochoa of *polynucleotide phosphorylase*, an enzyme which in the cell degrades RNA but which, in the test-tube, can be 'forced' to catalyse the reverse reaction and synthesize RNA. This reaction does not require a DNA template and is unrelated to transcription.

To use an artificial RNA molecule as a message for translation, a cell extract able to synthesize proteins is needed. Such a *cell-free*, or *in vitro*, system must contain all the cellular components necessary for protein synthesis but must lack endogenous mRNA so that protein synthesis occurs only when the artificial message is added. Marshall Nirenberg and Heinrich Matthaei at the National Institutes of Health Laboratories in the USA eventually perfected a cell-free system, prepared from *E. coli*, able to synthesize polypeptides specific to an added artificial RNA message.

According to scientific legend, Matthaei worked through the night of Friday 26 May and around 6 am the following morning discovered that when a homopolymer made up entirely of uridine nucleotides—poly(U)—was added to the cell-free system, polyphenylalanine was synthesized. The first codon could be assigned: 5'-UUU-3' for phenylalanine. Poly(U) was used in this first experiment because homopolymeric RNAs are relatively simple to synthesize. If the reaction mixture from which polynucleotide phosphorylase builds up the RNA molecule contains only uridine mononucleotides then only poly(U) will be made.

The three other RNA homopolymers were also synthesized, added individually to samples of the cell-free system, and the assignations AAA = lysine and CCC = proline were made. For reasons that have never been entirely clear poly(G) would not work and so the amino acid coded by GGG could not at this stage be identified. The construct heteropolymers, artificial RNA molecules containing more than just one nucleotide. This can be achieved by polymerizing mixtures of nucleotides with polynucleotide phosphorylase. There is a problem, however, as the random nature of the polymerization means that the absolute sequence of the resulting RNA molecule is not known. For instance, a random heteropolymer of A and C contains eight different codons (AAA, AAC, ACA, CAA, ACC, CCA, CAC, CCC), in any order, and codes for a polypeptide containing six different amino acids (proline, histidine, threonine, asparagine, glutamine and lysine).

Clearly these six amino acids are coded by the eight possible codons, but which is coded by which? An answer is at least partially provided by using a known amount of each nucleotide in the reaction mixture for RNA synthesis: For example, if the ratio of C to A is 5:1 then the probability of a CCC codon occurring is much higher than the probability of an AAA, In fact the frequencies of each of the eight possible codons can be worked out and compared with the amounts of each amino acid in the resulting polypeptide. The codon allocations provided by this method are not definite but are statistically probable and can be cross-checked by changing the composition of the heteropolymer.

These techniques allowed Nirenberg, Matthaei and Ochoa to propose meanings for most of the 64 codons of the genetic code. Although homopolymers and random heteropolymers allowed most of the genetic code to be worked out, unambiguous determination of each codon required two additional types of experiment. Ordered heteropolymers were synthesized by H. Gobind Khorana. These are made by polymerizing not mononucleotides, but known dinucleotides such as AC (producing ACACACAC, which contains two codons, ACA and CAC) or trinucleotides such as UGU (producing UGUUGUUGU, codons UGU, GUU, UUG). The simpler polynucleotides produced by these messages allowed the meaning of several difficult codons to be determined.

The triplet binding assay was developed by Nirenberg and Philip Leder in modification of the cell-free protein synthesizing system. It was discovered that purified ribosomes attach to an mRNA molecule of only three nucleotides (a single codon, in fact) and bind the correct amino acid-tRNA molecule. As triplets of known sequence could be constructed in the laboratory, it was possible with the binding assay to check the previously assigned codons and to allocate virtually all of the remaining ones. All amino acids except methionine and tryptophan have more than one codon, so that all the possible triplets have a meaning, despite there being 64 triplets and only 20 amino acids.

Chapter 4

Vertebrate Physiology

Respiratory mechanisms

In vertebrates special surface areas are provided in the gills or lungs where the circulating blood can pass within three microns of the environmental medium; comparable distances exist between the blood and each internal tissue cell. Ventilation of the gills with water or of the lungs with air provides a constantly changing environmental medium adjacent to these surfaces. Circulation of the blood continually brims blood that is poor in oxygen and rich in carbon dioxide adjacent to the same respiratory surfaces. Thus, in the gills or lungs, steep diffusion gradients are constantly maintained. This favours the rapid diffusion of carbon dioxide outward and oxygen inward so that the blood leaves the respiratory surface rich in oxygen and poor in carbon dioxide.

Similar physical relationships favor efficient gas exchanges between the blood and the metabolizing tissue cells deeper within the animal body. The process of ventilating the gills with water or the lungs with air and the diffusion exchanges of gases between these surfaces and the blood constitute what is known as external respiration. The gas exchanges between the blood and the metabolizing tissue cells and the utilization of oxygen and production of carbon dioxide by those cells constitute the process of internal respiration. All other factors being equal, the volume of metabolizing tissue determines the oxygen demand of the animal. This entire demand must be met by gas exchange through the available surface area of the gills or lungs. Thus, where the demand is greater, the gill or lung area must also be greater. As an organism grows larger, its volume increases approximately as a cube, whereas its total surface area increases only as a square. Even if the entire surface area of the animal could function in respiratory exchanges with the environment, this would be insufficient to meet the oxygen demands of larger animals.

The general surface (skin) of a vertebrate animal is not satisfactory for respiratory exchanges. It cannot be thin enough to permit efficient gas diffusion through it and still be tough enough to provide a protective covering for the animal. This is why special respiratory surfaces with greatly increased areas for exchange must be provided in the form of gills or lungs. The total respiratory area of human lungs for example, about 40 times as great as the total skin area! If the adjacent environmental medium remained stationary, it would soon be depleted of its oxygen and saturated with carbon dioxide. Consequently, it must be constantly changed.

Respiration in fishes

Vertebrates either pump water past the gills or they repeatedly fill and empty the lungs with new air. Both processes are known as ventilation. Similarly, if the circulating blood remained stationary, that portion adjacent to the respiratory surfaces would soon become saturated with oxygen and unable to accept more. It would also become thoroughly depleted of carbon dioxide so that no further diffusion exchange of gases could occur. The blood must continuously circulate past the respiratory surfaces in order to maintain continuous gas exchanges by diffusion. Within limits, increases in either the ventilation rate or the circulation rate will result in increased rates of gas exchange with the environment. Both these mechanisms are used by vertebrates to increase the availability of oxygen to their tissues during exercise or other periods of heightened oxygen demand. It is a matter of common experience that when exercising we breathe more deeply and more rapidly and our heart rates increase. Oxygen diffuses through air more than 3 million times faster than it diffuses through water. Moreover, the solubility of oxygen in water is so low that water contains only one twentieth as much oxygen as does air. This means that oxygen is far less available to aquatic organisms than it is to air-breathers. This is a limiting factor even to air-breathers, however, since oxygen must be dissolved in the fluids of the respiratory membrane and in the blood itself during the process of diffusion. Increased salinity and temperature each reduces the oxygen concentration of water. Thus, marine fishes and fishes living in warm tropical waters have less oxygen available to them than do freshwater fishes. Moreover, increased temperature results in an increase in the metabolic rate of cold-blooded vertebrates.

In warmer waters, then, the oxygen demand of a fish is increased at the same time and by the same factors that decrease the availability of oxygen. Some bodies of water are extremely poor in oxygen because of poor circulation of the water or because oxygen-consuming processes, like the rotting of submerged vegetable matter, are present. The decay of plant materials also produces carbon dioxide, and waters in which such decay occurs are generally rich in that gas. In such bodies of water only the immediate surface may contain sufficient oxygen to meet the needs of aquatic vertebrates. All of the factors mentioned thus far in this section make it clear that there are many advantages to breathing air instead of water. Thus, the development of air-breathing apparatus was favoured long before terrestrial vertebrates evolved. Although air is a better source of oxygen than water, the terrestrial environment is more demanding in other ways. Air is desiccating and far more susceptible to variations in temperature and other physical conditions.

The transition from aquatic to terrestrial life would seem, therefore, to have been an attractive move but a difficult one to achieve. The transition from aquatic to terrestrial life may well have been preceded by the adaptation of certain fishes to oxygen-poor waters. In such situations many modern fishes rise regularly to the surface to gulp in air. This behaviour is frequently seen, for example, in ordinary goldfish kept in an unaerated aquarium.

Vascular membranes of the pharynx, esophagus, stomach, and even of the intestine serve as accessory respiratory surfaces for such air breathing. Certainly any increase in the area of respiratory surface would greatly benefit a fish, and thus the development of an esophageal diverticulum, such as a swim bladder or a lung, would be favoured. This is apparently the adaptation that actually appeared in more than a single line of descent, including the ancestors of the modern lungfishes.

The Dipnoi arose in the tropical swamps characteristic of the Devonian period, a little more than 400 million years ago. The warm climate and abundant vegetation in these swamps must have greatly reduced the oxygen content of the water. Moreover, periodic droughts occurred, and portions of the swamps must have dried up, leaving only scattered pools. The animal that could breathe air could survive either by migrating overland to remaining ponds or by aestivating until the next rainy season. Because the Dipnoi possessed lungs in addition to gills and because they were able to reduce their rate of metabolism in order to survive long periods of aestivation, the lungfishes were able to employ successfully both of these survival measures. Of the lungfishes living today, Neoceratodus (Australia) has well-developed gills and only a single, poorly developed lung; it cannot remain long out of water. Protopterus (Africa) has paired lungs and less well-developed gills; it can get along about equally well in either water or air. Lepidosiren has highly developed, paired lungs, but the gills are so poorly developed that the animal will drown it held under water! The three living groups, then, illustrate three stages in the probable evolutionary conversion from aquatic to aerial respiration. It is not sufficient for an animal merely to have lungs in order to breathe air.

Elimination of carbon dioxide by animals is not ordinarily a limiting factor even in water. Carbon dioxide is 28 times more soluble in water than is oxygen. Moreover, when dissolved in water carbon dioxide forms a hydrate, carbonic acid (H_2CO_3). This dissociates to form hydrogen and bicarbonate ions, which often react with other dissolved substances to form various carbonates and bicarbonates. Such reactions reduce the concentration of carbon dioxide, as such, and this helps to maintain steep diffusion gradients between the animal surface and the immediately surrounding environment.

Quite likely the earliest aquatic vertebrates had external gills much like those of modem tadpoles. These are filamentous structures that hang outside the pharyngeal region and are ventilated by the fish's movements through the water. Such external gills, however, are subject to predation and other injuries. Their gas exchange efficiency is much reduced when the animal is not swimming, and they interfere with the fish's streamlining when it is swimming.

The development of internal gills, in gill pouches or in opercular cavities, removed many of these disadvantages. The overlying operculum of typical fishes provides protection for the gills and communication between the gills and the oral-pharyngeal cavity provides a means of pumping water over them even when the animal is at rest. Branchial arches bridge across the opercular cavity, which opens between the oral-pharyngeal cavity and the outside gill opening. These branchial arches bear successive gill filaments, or primary lamellae that appear as flat plates lying parallel with each other. Each primary lamella, in turn, bears smaller secondary lamellae at right angles to itself. Together these lamellae form a sievelike structure through which water passes. Blood capillaries in the lamellae pass within 0.5 to 0.3 micron of the water-bathed surface, so diffusion distances between water and blood are very short. Blood flow through the secondary lamellae is in a direction opposite to that of the passing water, and this greatly increases the efficiency of gas exchange. If water and blood passed adjacent to each other in the same direction, a given "drop" of water and "drop" of blood would be contiguous longer than necessary to complete all of the possible diffusion of gases between them.

During this countercurrent exchange, however, each "drop" of water is constantly brought adjacent to different "drops" of blood. Experimentally it has been found that reversing the direction of water flow through the gills of a fish reduces the oxygen uptake to about one eighth of its normal value simply because, under these conditions, blood and water are traveling in the same direction. The primary and secondary lamellae of typical fish gills form a meshwork or sieve having up to a quarter of a million pores. This provides a total surface area for gas exchange that is about ten times the fish's skin area. Because the pores are very small, all of the water passing through them passes very close to the respiratory surfaces of the gill lamellae. As a result, about 80 per cent of the oxygen available in the water is taken tip by the blood passing through the gills. The ventilation of the gills of a teleost fish (bony fish) is accomplished by means of a double pump. In the first phase of pumping, the mouth opens and water is drawn into the oral-pharyngeal cavity when the floor of this cavity is depressed. The operculum remains closed during this phase, but the opercular (gill) cavity beneath it is actively expanded. This draws water from the oral-pharyngeal cavity through the gills and into the opercular cavity.

In the second phase of pumping, the mouth closes and the opercular cover opens. The oral-pharyngeal floor is raised again, driving more water through the gills and back to the outside through the opercular cover opening. This sequence of events, involving as it does both an oral-pharyngeal pump and an opercular cavity pump, results in an almost continuous flow of water through the gills. Intermittent flow, such as would result from the action of a single pump, of course would mean reduced efficiency of gas exchange.

In the elasmobranch fishes the process of gill ventilation is similar except that water enters the oral-pharyngeal cavity through a special spiracle, located on each side of the head, rather than through the mouth. Also, instead of being located in a single opercular cavity, the gills are located in each of several gill pouches that open independently onto the surface. The spiracle is actually a modified first gill pouch which contains no gills. The others each contain a single branchial arch and attached lamellae. The substitution of a spiracle for the mouth in the process of ventilation is probably related to the fact that many sharks and rays are bottom feeders indeed, all the ancestral elasmobranchs were and water entering the mouth would be contaminated with bottom detritus. Both teleosts and elasmobranchs sometimes suspend active water pumping when they are swimming, since the forward motion forces a sufficient amount of water over the gills through the opened mouth.

Some fishes can never stop swimming forward because they cannot achieve sufficient ventilation by oral-pharyngeal and opercular pumping alone. Many, but not all, teleost fishes possess a gas-filled bladder that arises during embryonic development as a diverticulum of the digestive tract. These bladders bear a superficial resemblance to a simple lung. In other species the bladder loses its connection with the digestive tract but remains filled with gas. In both cases it is possible for the fish to inflate the bladder by secreting into it oxygen from the blood.

The structure of the swim bladder is fairly uniform in all species. It is an oblong structure that has a smaller side chamber, known as the oval, communicating with it. Often muscles in the wall are present that can close off the opening between the swim bladder proper and the chamber of the oval. In the walls are two vascular areas; the gas gland, which actively inflates the swim bladder, and all area in the wall of the oval through which oxygen call be reabsorbed to deflate the bladder. The relative amounts of blood passing through these two areas determine the degree of inflation at any given time.

Moreover, closing off the opening into the oval completely prevents gas reabsorption and, thus, favours inflation of the bladder. To transport oxygen from the blood into the swim bladder means moving it against a steep concentration gradient, since the content of the swim bladder is usually pure oxygen. The mechanisms involved ill this process are not yet completely understood but enough is known to facilitate some conjecture. The gas gland consists of a network of blood vessels arranged so that each makes a hairpin turn near the lining of the swim bladder. Blood flows into each hairpin loop and gives up some of its oxygen before flowing back out through the efferent limb of the loop. This oxygen passes into the swim bladder.

During active inflation of the swim bladder the metabolism of glucose increases markedly in the region of the hairpin turns, and this results in an accumulation there of carbon dioxide and lactic acid. These substances both lower the ability of the blood to retain oxygen, and consequently it is released from the blood in that region. The swim bladder primarily serves to adjust the buoyancy of the fish so that it may remain at a particular depth without constantly needing to overcome any, natural tendency to either rise or sink.

The density of a fish averages about 1.076, so without such a gas-filled bladder it would tend to sink slowly. It has been calculated that in fresh water a fish weighing 100 grams should require a swim bladder with a capacity of about 7 milliliters to provide the necessary buyoyancy to offset the fish's density. Since sea water has a greater specific gravity, 1 swim bladder of only a 5-milliliter capacity would be adequate to accomplish the same purpose. These are approximately the sizes of the swim bladders actually found in such fishes. Teleosts with heavy skeletal structures and those inhabiting fresh water are most benefitted by swim bladders, and it is in these fish that the organ most often is found.

Marine fishes with lighter skeletons are less often so equipped. Elasmobranch fishes do not have swim bladders; instead these animals store low-specific gravity fats, especially in the liver, to achieve the necessary buoyancy. Swim bladders have a disadvantage that is not shared by the fat-storage buoyancy system of elasmobranchs. Just as the bubbles in a glass of soda increase in size and in rate of ascent as they rise, so the swim bladder expands and becomes increasingly buoyant as the fish approaches the surface of the water. This can be offset only by reabsorption of oxygen from the swim bladder as the fish rises toward the surface. It has been calculated that fish at a depth of 20 meters could rise rapidly to a depth of 15 meters without difficulty. Swim bladders also sometimes function in connection with hearing and with sound production or amplification in various fishes.

Some authorities believe that the evolution of the swim bladder has been homologous with the development of lungs; others disagree with this contention. The argument against homology is based upon the fact that the swim bladder is typically a dorsal out pocketing from the gut, is a single, unpaired structure, and is supplied by blood circulation in series with the general body circulation. The arteries supplying it, moreover, are derived from the arteries supplying the gut in other vertebrates. Lungs are typically paired outpocketings from the ventral side of the gut and are supplied by arteries that are in parallel with the general body circulation.

Respiration in amphibians

All amphibian larvae are aquatic and equipped with external gills. Skin respiration, especially in the long, thin tails of tadpoles, is probably an important adjunct to gill respiration in these larvae. The lungs of adult amphibians are, like those of lungfishes, relatively simple, uncomplicated sacs without internal subdivisions. In this feature they resemble swim bladders more closely than they resemble typical vertebrate lungs. Evolution of lungs has led to greater and greater degrees of internal subdivision and thus to greater total surface areas for the exchange of gases with the blood. The ventilation of amphibian lungs depends upon an oral-pharyngeal pump reminiscent of that used by teleost fishes to ventilate the gills. Air is drawn in through the nostrils rather than through the mouth, but this is accomplished by depression of the oral-pharyngeal floor. The nostrils then close and the oral-pharyngeal floor is again raised driving the air through the open glottis into the lungs. The glottis at the entrance to the trachea then closes, holding the air in the lungs for a time. This allows the air in the lungs to be more thoroughly depleted of its oxygen, since turbulence there brings more of the air into direct contact with the lung surfaces. While the air is being held in the lungs, there is a repeated oral-pharyngeal pumping of air in and out through the nostrils. Authorities differ in their opinions of the purposes served by this. Probably it is purely olfactory behavior but it may also serve a significant respiratory role. The lining membranes of the mouth and pharynx are highly vascular, so some gas exchange with the blood undoubtedly occurs. Finally, the glottis reopens and the lungs are emptied by contraction of smooth muscles in their walls. Skeletal muscles in the thoracic walls may also aid in emptying the lungs.

Respiration reptiles

Mst reptiles are inflated by active expansion of the thoracic walls while both the nostrils and the glottis are held open. After lung inflation, the glottis is closed, and the muscles of the thoracic walls relax again. This places the air in the lungs under some positive pressure, which may increase the rate of oxygen extraction from it. In any case, holding the air for a time presumably favors more complete oxygen extraction, as it does in amphibian lung ventilation. Also, as is true in the amphibians, repeated oral-pharyngeal pumping of air occurs while the glottis is closed. Again, this may be simply olfactory activity or it may contribute to respiratory gas exchange through the oral and pharyngeal walls. The glottis is reopened, and the lungs are emptied by the movements of smooth muscles in the walls of the lungs and, possibly, by the contraction of skeletal muscles of the thoracic walls. This represents an improvement over the amphibian mechanism for lung ventilation since active thoracic expansion, rather than oral-pharyngeal pumping, is used to inflate the lungs.

The oral-pharyngeal pump still operates while air is being held in the lungs, however. Also, the oral-pharyngeal pump can be employed by some reptiles to super-inflate the lungs. Following normal inhalation oral-pharyngeal activity can be used to drive larger quantities of air into the lungs. In these animals the lung cavities frequently communicate with extensive, non-respiratory air sacs that can be filled by this means. Superinflation causes the entire animal to puff up to a larger size, and this behavior is used to frighten predators, in mating displays, and to wedge the animal into holes or cracks so that it cannot easily be captured. The possession of the accessory air sacs in certain reptiles preceded in evolution the very important air sacs of birds. Because the major portion of the body wall of turtles is rigid, general expansion and contraction of the thoracic walls cannot occur. In these animals there are specially adapted skeletal muscles, lying adjacent to the limbs, that produce movements of the soft parts of the thoracic wall to draw air into the lungs. Probably movements of the limbs themselves also aid in ventilation. There is evidence to suggest that the muscles involved are derived from the intercostal muscles and diaphragm; these are the same muscles used for respiration in other reptiles and higher vertebrates.

Respiration in birds

Lungs and extensive non-respiratory air sacs of birds are both ventilated by active expansions and contractions of the thoracic and abdominal walls. Oral-pharyngeal pumping, like that seen in fishes, amphibians, and reptiles, is not present. During flight the respiratory movements may be correlated with the wingbeats so that the same muscular movements can accomplish both tasks. Inspired air is not held for a time as in amphibians and reptiles but is, instead, moved rhythmically in and out of the lungs. Except for the birds and, to a lesser extent, some reptiles, the lungs of vertebrates are cul-dc sacs, i.e., blind pockets connected with the outside atmosphere through the trachea and other respiratory passages. In such lungs even the most forceful expiration does not completely empty the system of air. The respiratory passages remain filled, and with the next inspiration this "dead air" is drawn back into the lungs. Fresh air from the outside, thus, is mixed with air that has not been exhaled at all. At best, the air in the lungs is a mixture of fresh air and dead air from which most of the oxygen had already been removed. The obvious inefficiency of such a system is corrected in the birds.

During inspiration the dead air is drawn through the lungs and into the air sacs. The lungs are then filled with the incoming fresh air. Without this increase in respiratory efficiency it would be difficult, if not impossible, for birds to meet the great oxygen demands imposed by the activity of flying. To do so would require that the lungs be very much larger. Actually, the benefits acquired with the addition of air sacs are so great that bird lungs are, in general, comparatively smaller than those of other vertebrates. Besides improving respiratory efficiency, the air sac system probably also functions in the control of body temperature. It has been calculated that for cooling purposes a pigeon in active flight would require a volume of air three times as great as that required for respiratory gas exchange. The air sac system ramifies throughout the body, and undoubtedly ventilation of this system effectively carries away much of the heat generated by active flight muscles. The air sacs, occupying as they do a substantial percentage of the total volume of the bird, reduce its weight, and, of course, this is another important contribution to flight. Because the air sacs remove the problem of dead air space in the respiratory passages, the volume of the dead air space is no longer a critical factor; it can be much larger without interfering with efficient gas exchange in the lungs. Frequently the tracheas of birds are doubled upon themselves lengthwise and they often contain large, resonating vocal chambers. By comparison, even the addition of it small amMuch more difficult. This is a familiar fact to anyone who has made use of a snorkel in skin diving.

Since the snorkel is effectively an extension of the respiratory passage, it, therefore, adds its own volume to the dead air space. It is also familiar to any boy who has attempted to emulate the heros of western books and movies by trying to breath through a hollow reed while hiding submerged in the river. The internal structure of bird lungs is quite different from that of most other vertebrates. Air is drawn from the trachea through a main bronchus and into the air sacs.

Smaller tubules called parabronchi branch off from the main bronchus, and still smaller air capillaries branch off from the parabronchi. Presumably air currents and, perhaps, physical movements of the lungs ventilate all of these passages while air is passing through the main bronchus. Ventilation of the air capillaries occurs, then, both during inspiration and expiration. This undoubtedly results in a very complete extraction of the oxygen contained in the air.

Blood circulatory mechanisms

Circulation in fishes

The typical circulatory system in the fish consists of a series of closed vessels much like those of all higher vertebrates. Consequently, a single heart is able to maintain circulation at increased Pressures, and thus greater activity of the animal is made possible. The fish heart has a single, undivided atrium and a single, undivided ventricle. In addition, it has two other chambers arranged in series with the atrium and ventricle. These chambers are lacking in mammalian hearts. Venous blood is collected in a sinus venosus, which contains pacemaker tissue. This chamber empties into the single atrium and probably plays a role similar to that played by the mammalian atrium; that is, it guarantees a continuous, uninterrupted influx of venous blood to the heart. Blood passes from the atrium into the single ventricle and then into a final chamber, the bulbus arteriosus. This chamber probably corresponds to the first portion of the mammalian aorta. It is equipped with several valves and apparently serves to smooth out the arterial pressure waves. Such an initial dampening of pulse waves produces a more uniform, continuous flow of blood through the gills.

It has been suggested that the very low pressures of venous blood would be insufficient to inflate a heart chamber unless that chamber bad very thin walls like the sinus venosus. The sinus venosus may also serve, then, to these the blood pressure sufficiently to inflate the thicker-walled atrium. Similarly, the atrium can in ate the even thicker walled ventricle. Each chamber may act in sequence to boost the blood pressure by degrees from venous to arterial levels. Fish hearts receive parasympathetic nerves from the tenth cranial nerve (vagus), but there is no evidence of any corresponding sympathetic innervation. The stroke rate is not increased by increasing physical activity. Activity does increase the venous return by milking action, however, and this results in increased cardiac output due to the Starling effect. Also, the stroke rate varies to some extent with changes in the environmental temperature. Since the metabolic demand for oxygen and nutrients also increases with temperature in cold-blooded animals, this response of the heart is of importance. An interesting adaptation appears in elasmobranch fishes (sharks, rays), in the electric eel (Electrophorus), and perhaps in some other fishes as well. During inflation of the pharynx with water, pressure receptors in the walls of the pharynx are stimulated. Nerve impulses from these receptors to controlling centers of the brain produce an inhibition of the heart rate. As a result, cardiac filling is prolonged and increased.

The next heart beat, because of the Starling effect, is a more forceful one and the stroke volume of that beat is increased. Water and blood then pass over the gills simultaneously and the efficiency of gas exchange is improved. A limiting factor in blood circulation in fish is the fact that the blood goes from the heart to the gills and then ton into the systemic arteries. The pressure of blood entering the gills ranges from 15 to 70 millimeters of mercury in different fishes and may be as high as 120 millimeters of mercury in particularly active species such as salmon. The pressure drop imposed by the gill capillaries, however, may amount to as much as 20 to 40 millimeters of mercury. Thus, while the gills receive efficient, high-pressure circulation, the rest of the animal must get along on reduced blood pressures.

If there were some sort of booster pump following the gills, this limiting factor would be removed and the entire body would receive a high-pressure circulation. Although no such accessory heart ever appeared in fishes, the further evolution of the heart in higher vertebrates has accomplished the same thing. The pressure drop across the gills is probably of much less importance in the Dipnoi (lungfishes) since some of the branchial arches of these fishes lack gills. Thus, the total gill resistance is not as great as in other fishes. The major evolutionary advance seen in the lungfishes, however, is the addition of lungs as accessory respiratory organs. These lungs receive blood that has already passed through the gills so they can only add to whatever oxygenation of the blood has already taken place in the gills. Nevertheless, this gives the animal the opportunity to supplement the oxygen available from a stagnant, aquatic environment It also increases the ability of the animal to choose between an aquatic or aerial form of respiration, at least for limited periods of time.The atrium of the lungfish heart is divided into right and left chambers and the ventricle is also partially divided. Blood from the lungs enters the left atrium and then the left half of the ventricle.

Despite the fact that the ventricle is incompletely divided, the partial division tends to keep this blood on the left side so that it is ejected into the aorta on its left side. An incompletely developed spiral valve (partition) of the aorta then shunts this oxygenated blood, which came from the lungs, into the first two branchial (gill) arches. These two arches carry blood mainly toward the head. Thus, the brain is guaranteed a sufficient oxygen supply. Similarly, blood from the general body circulation enters the right atrium of the lungfish. This relatively unoxygenated blood passes into the right half of the ventricle and is shunted by the spiral valve of the aorta into more posterior gill arches. From these it passes mainly to the lungs. Thus, the unoxygenated blood is guaranteed good aeration in both gills and lungs.

Circulation in amphibians

Circulatory systems of adult amphibians are somewhat variable since some species are completely aquatic and have gills while others are largely terrestrial and more dependent upon lungs. The larvae of all amphibians are aquatic, of course, and they have circulatory systems similar to those of fishes. The Anura lose the gills at the time of metamorphosis and replace them with lungs. The adult circulatory patterns of frogs and toads are similar to those of lungfishes except, of course, that the gill circulation is absent.

The amphibian skin also serves as an accessory respiratory surface, and the pulmonary artery has cutaneous as well as pulmonary branches for this reason. The atrium of the frog heart is completely divided into right and left chambers but the ventricle is undivided. The spiral valve of the aorta is better developed than in the lungfishes, and it is probable that only minor mixing of blood from the right atrium and that from the left atrium occurs in the ventricle. At least this has been verified in those species which have been thoroughly studied.

Oxygenerated blood from the lungs enters the left side of the heart and goes primarily into the general systemic circuit. Unoxygenated blood returning from the general body enters the right side the heart and is pumped primarily to the lungs and skin. In the amphibian heart the sinus venosus is also present and serves, as in the fish heart, to receive blood From the venous drainage and to act as an initial pressure booster. Even though the sinus venosus is not divided, only minimal mixing of oxygenated and unoxygenated blood occurs there. The amphibian heart also contains a bulbus arteriosus, as does the fish heart. It serves, as in fishes, to smooth out the arterial pressure wave. The heart is innervated by both sympathetic and parasympathetic nerves, which function in pressure control just as they do in mammals.

Circulation in reptiles

Non-crocodilian reptiles have circulatory patterns and hearts similar to those of frogs and toads except that in these reptiles the skin plays no role in respiratory gas exchange. Also, the ventricle of the heart is partially divided like that of the lungfishes. In the crocodilian reptiles both the atrium and the ventricle are completely divided so that no mixing of oxygenated and unoxygenated blood occurs. Moreover the spiral valve of the aorta is so well developed that the aorta is divided to form essentially an independent aorta and Pulmonary artery.

Circulation in birds

The circulatory pattern of birds is essentially like that of mammals except for minor details that are of no physiological significance. For example, mammals retain the left aortic arch and lose the right one, whereas birds retain the right aortic arch and lose the left. Thus far, research has shown that the controlling mechanisms of blood pressure in birds are like those of mammals. The heart rate of birds is greater and the arterial blood pressure is higher than in mammals of comparable size. The bird heart is proportionately larger, but the demands of flight are so great that it must also beat faster. It would be impractical to increase the minute-volume of the bird heart by developing a still larger heart since this would also increase the "payload" to be carried in flight. Yet, increased blood pressure is necessary to meet the rigorous requirements of flying. Thus, the heart and blood vessels are operating, at near maximum to sustain everyday activity.

Circulation in mammals

The mammalian heart is a double pump, each half of it serving a separate portion of the circulation. The right half receives blood from the general systemic circuit and pumps it to the lungs. The left half accepts blood returning from the lungs and pumps it into the general systemic circuit again. The two halves of the heart are each, in reality, separate pumps but they operate in unison with each other. Blood from the systemic circuit enters the right atrium of the heart and blood from the lungs enters the left atrium at the same time. From the atria the blood flows into the corresponding ventricles on the two sides of the heart. The ventricles contract in unison expelling the blood into the arteries; blood from the right ventricle flows into the pulmonary artery leading to the lungs and blood from the left ventricle enters the aorta. Blood from the lungs returns to the left atrium through the pulmonary veins and blood from the general systemic circuit returns to the right atrium by way of the venae cavae. Valves within the heart prevent the blood from flowing through it in the opposite direction.

Necessarily, the volume of blood pumped by the right side of the heart must equal, over any short period of time, the volume pumped by the left side of the heart. This is necessary if both the pulmonary and the systemic circuits are to remain sufficiently but not overly filled. If the right side of the heart pumped a larger volume than the left side, the pulmonary circuit would become overfilled and the systemic circuit would be depleted of its normal content of blood. This would result in increased pulmonary blood pressure and decreased systemic blood pressure. Since the pulmonary circuit is much shorter than the systemic circuit, its normal content of blood is smaller in volume.

Moreover, since this imposes less frictional resistance to blood flow, there is less pressure drop through the pulmonary circuit than through the systemic circuit. This means that the right side of the heart need not be as muscular or develop as much pumping force as the left side. This difference is reflected in the comparative thickness of the right and left ventricular walls and in other details of their structure. The contraction of the chambers of the heart is called systole and relaxation is called diastole.

During systole of the ventricles, blood is ejected into the pulmonary artery and the aorta, and the blood pressure in those arteries rises to a maximum. The maximum arterial pressure is therefore spoken of as the systolic pressure. While the ventricles are relaxed and are being refilled with blood from the atria, the pressure in the arteries falls as a result of the run-off of blood into the capillaries and veins. Arterial blood pressure thus decreases to a minimum just before the next ventricular systole. This minimum arterial pressure is called the diastolic pressure because it coincides particular diastole. Each side of the heart has a valve between the atrium and the ventricle known as the atrioventricular valve.

The atrioventricular valves of the two sides of the heart are similar in structure except that the right valve has three cusps or flaps while the left valve has only two. The right valve is, accordingly, called the tricuspid valve and the left valve is called the bicuspid valve. A more commonly used name for the left atrioventricular valve is the mitral valve. This name comes from the supposed resemblance of this valve to the miter worn by the bishops of liturgical Christian churches. Both atrioventricular valves are somewhat funnel-shaped structures.

The free edges of their cusps hang downward into the ventricular cavities and are anchored by tendon-like strands of connective tissue known as chordae tendineae. At the point of attachment of each of the chordae tendineae to the ventricular wall there is a small hillock of muscle called a papillary muscle. As the ventricles contract in systole, increasing pressure within them forces the cusps of the atrioventricular valves together. This closes these valves and prevents the backflow of blood into the atria. If it were not for the presence of the chordae tendineae anchoring the valve cusps to the sides of the ventricle walls, the pressures developed in the ventricles would be sufficient to turn the atrioventricular valves inside out and evert them into the atrial cavities. During ventricular contraction, the ventricles shorten and their diameters are decreased. The simultaneous contraction of the papillary muscles prevents any slack from appearing in the chordae tendineae as a result of ventricular shortening. Otherwise such slack might be sufficient to allow the valves to be everted into the atrial chambers. When the ventricles relax, pressure within them decreases, and the atrioventricular valves are allowed to open again. This permits blood to flow into the ventricles from the atria and from the venae cavae and pulmonary veins.

The pulmonary artery and the aorta each have a valve at the point of exit from the corresponding ventricles. These are the semilunar valves and they are quite similar in both eases, but they are totally unlike the atrioventricular valves in structure. Each semilunar valve is composed of three pocket-like cusps with the pocket opening directed away from the ventricular chamber. These pockets fold outward against the arterial walls when the valves are opened and they fit neatly together to block the lumen when the valves are closed. It is interesting to note in passing that a bicuspid valve of similar design would close just as well but could not be opened as completely. A tetracuspid valve of the same type would only add further complexity without materially improving the opening or closure. Three cusps of this type appears to be the ideal arrangement.

When the ventricles contract in systole, the increasing pressure within them forces the cusps of the semilunar valves outward and against the arterial walls so that blood can be ejected into the pulmonary artery and aorta. When the ventricles relax in diastole, the pressure within the ventricles falls, and as soon as it is less than the pressure within the arteries, the backwash of arterial blood attempting to reenter the ventricles catches in the pockets of the valves and snaps them shut. It is important to recognize that all the valves of the heart are opened and closed by the differential fluid pressures on their two sides;

they are not muscular structures capable of independent action.

Excretory mechanisms

The vertebrate nephron apparently did not evolve directly from any known excretory organ found in modern invertebrates. However, similar chemical and physical principles operate in the excretory organs of both vertebrates and invertebrates. Adult Myxine hagfishes have an excretory system that superficially resembles the metanephridia of invertebrates, i.e., simple tubules draining celomic fluids through the body wall to the outsid. Nearby, in the celomic wall, there is a glomerulus-like capillary bed that, by forced filtration, increases the rate of formation of celomic fluid. The nephron tubule, then, reabsorbs some substances from the fluid as it drains toward the outside of the animal body.

Excretion in fishes and amphibians

In some sharks and amphibians the communication of the nephron tubule with the celomic cavity persists, but there is a more intimate connection between the tubule and the glomerulus—a connection resembling Bowman's capsule. In higher vertebrates the direct connection with the celomic cavity is lost. The tissue fluids of some marine invertebrates have the same total osmotic concentration as that of the surrounding sea water and there is no tendency to gain or to lose water by osmotic flow in these animals. The specific solutes responsible for internal osmotic pressure are never of the same concentration as those outside the animal, however

Therefore, mechanisms to prevent solute diffusion through the body surface must be present. Moreover, even, organism must be able to accumulate specific molecules necessary for its metabolism and to retain them in the body in greater concentration than in the surrounding medium. The accumulation of such materials and the excretion of wastes demands that the animal have surface membranes that are permeable to solutes, and the movement of such solutes through these membranes results in the simultaneous movement of water as well. Finally, all known animal cells maintain electrical charge differences on the two sides of the cell membrane by maintaining differential concentrations of various ions on the two sides. With all of these facts in mind, the inescapable conclusion is that active transport of substances through membranes between internal fluids and the surrounding medium must have appeared as early as life itself.

A fresh-water vertebrate has an internal salt concentration that is higher than that of the environment. Consequently, there is a tendency to lose salts by diffusion and to gain water by osmotic uptake. If mechanisms must be present to prevent or to compensate for these tendencies. One might suppose that such an animal could simply develop a surface covering that would be impermeable to either water or salts. Such impermeability, however, would also prevent the vital exchange of respiratory gases and the accumulation of other specific metabolic requirements. The problem can be minimized, of course, by limiting the extent of such permeable surface membranes, and such is the case for most vertebrate animals. Minimizing a problem does not solve it, so there is still an unavoidable salt loss and water gain through intestinal, renal, and respiratory membranes, and these losses and gains must be compensated for by other mechanisms.

The greatest single factor in this compensation is the presence of a kidney which is capable of retaining salts and excreting water. The fact that all vertebrates have kidneys which are fundamentally designed to accomplish these ends argues strongly for a fresh or brackish-water origin. It would be most difficult to explain the appearance of such a kidney in a hyperosmotic environment. The renal retention of salts is not, of course, 100 per cent efficient. Moreover, salt loss still occurs through intestinal and gill membranes as well. Some fresh-water fishes obtain sufficient quantities of salt in their diets to offset the unavoidable salt losses; others are capable of active, inward salt transport through the gill membranes. Even the freshest 4 natural waters contain some salts in solution and by active transport fishes can selectively absorb them from the environment. Fresh-water fishes drink little or no water, since the osmotic uptake through the gill membranes is already in excess of their requirements.

Marine teleost fishes have an internal salt concentration that is less than that of the surrounding environment so they tend to lose water by osmosis and to gain salts by diffusion through the permeable surface membranes. The possession of kidneys primarily designed to retain salts and to excrete water is a liability under these conditions. The number and size of the functioning nephrons and the minute-volume of blood supply to the nephrons are all reduced in marine teleost fishes. In a few cases glomeruli are completely lacking in the adults. The embryos of all fishes have glomeruli present at some stage of development, however, and this indicates probable descent from forms with typical vertebrate nephrons. It is interesting to note that in fishes such as salmon that migrate between fresh and marine waters, both kinds of salt and water problems must be dealt with.

In salmon the blood supply of the glomeruli is reduced when the animal enters sea water and is increased when it reenters fresh waters. Reduction of kidney function reduces the urinary water loss but it does not, of course, completely eliminate it. Moreover, substantial water loss also occurs through the gill membranes. Physiological mechanisms must be present, therefore, to compensate for these losses.

Unlike fresh-water fishes, teleosts in the sea drink large quantities 4 of water. The intestinal membranes absorb much of the water as well as the monovalent salts (Na^+, K^+) contained in it. Divalent salts (Ca^{++}, Mg^{++}, $SO_4^=$, $PO_4^=$) are absorbed in only very small quantities; most of them are passed, instead, with the feces. Those divalent salts which are absorbed are actively excreted by the kidneys. The absorbed monovalent salts are eliminated by active transport through the gill membranes. This active transport through gill membranes is the same as that encountered in fresh-water fishes, which use it to obtain salts from the environment, but here it operates in the opposite direction.

Fishes that migrate between fresh and marine waters (e.g., salmon) alter the degree and direction of active salt transport through the gill membranes as the salinity of the environment changes. In fresh water salts are accumulated from the environment and in marine waters salts are actively excreted through the same membranes. Thus, these "salt pumps" are constantly moving salts against the concentration gradient regardless of the direction of that gradient.

The marine elasmobranch fishes have developed mechanisms somewhat different from those just described for teleosts. These sharks, rays, and related forms also have typical vertebrate nephrons, indicating their probable descent from fresh or brackish-water vertebrates. Thus, they have kidneys which are primarily designed to retain salts and eliminate water, i.e., kidneys which are a liability in the marine environment. In addition, however, in the elasmobranch kidney their is a special segment of the nephron tubule which reabsorbs urea from the glomerular filtrate. Since urea is an osmotically active solute, the retention of urea in the blood effectively raises osmotic pressure of blood and tissue fluid. Elasmobranchs maintain an internal osmotic pressure that is higher than that of the surrounding sea water. This has required that elasmobranch tissues develop a tolerance to urea greater than that of-other vertebrate tissues. The result is that marine elasmobranchs, like fresh-water fishes, actually live in a hypoosmotic environment, in which the typical vertebrate nephron is appropriate. The internal osmotic pressure of elasmobranchs is due, in part, to urea rather than to other salts, so there is still a tendency toward the uptake of salts from the sea by diffusion through surface membranes. Salts accumulated in this way are actively excreted by a special cloacal gland, which does not occur in other vertebrates.

Most modern sharks and rays are marine, but the fossil record indicates that they were once more common in fresh or brackish waters. A few fresh-water sharks still exist but they are probably descended from marine ancestors since they too retain urea to some extent. In the fresh-water environment such urea retention is a liability because it raises blood osmotic pressure and, thus, magnifies the salt and water balance problems of a fresh-water vertebrate.

The tendency in these fresh-water elasmobranchs to lose salts by diffusion and to gain water by osmotic uptake is even greater than in the fresh-water teleost. Yet, the tolerance to urea has become, during evolution, a requirement. The elasmobranch heart, for example, will not beat normally in the absence of elevated urea levels. Thus, it has apparently been impossible for modern fresh-water sharks to completely eliminate urea retention. The amount of such retention has been greatly reduced, however, so this additional problem has been minimized.

The compensatory mechanisms of fresh-water elasmobranchs are generally the same as those of fresh-water teleosts. The kidneys produce an abundant, dilute urine while retaining salts. The special cloacal salt-excreting gland of marine elasmobranchs is, of course, absent in the fresh-water forms. Certain teleost fishes living in very cold surface waters of the Antarctic have blood osmotic pressures that are above those of the surrounding sea water. This is achieved by the retention of certain unidentified organic substances in a manner analogous to the retention of urea by elasmobranchs, However, these teleosts, derive an additional advantage from the depression of the freezing point of the body fluids. Among vertebrates, only the myxinoid cyclostomes (hagfishes) are, in a sense, osmoconformers. They have a total internal salt concentration equal to that of the surrounding sea water. Therefore, they are in osmotic balance with their environment.

The kinds and concentrations of individual specific ions are different of course, so individual ion regulation is still necessary. Selective intestinal absorption and renal retention accomplishes these requirements. This condition, much like that found in marine invertebrates, has been used to support the argument for a marine origin of the vertebrates. Certain amphibians, in brackish waters also tolerate increased internal salt concentrations, however, so the argument is not a conclusive one. Bufo viridis, the green toad of Europe and the Middle East, and Rana cancrivora, the crabeating frog of Southeast Asia, both retain urea and salts. Salt retention is the major factor in osmoregulation of the green toad, whereas urea retention is more important in the crab-eating frog. Also, it is more likely that the myxinoid fishes represent degenerate forms rather than ancestral species. The eggs of most aquatic vertebrates tend to conform with the environment, shrinking and swelling in response to changing osmotic forces of the medium until fertilization occurs. A decrease in the permeability of the egg membrane following fertilization isolates the internal fluids from the environment during development.

Excretion in terrostrial animals

The typical vertebrate kidney, designed as it is to excrete water, is something of a liability in the terrestrial environment and compensatory mechanisms are both physiological and behavioral. The Amphibia live at the junction between the fresh-water and terrestrial habitat. The larval period is always aquatic, but the adults of many species are largely, terrestrial. Consequently, adaptations to both habitats appear in this group of vertebrates. The water and salt balance mechanisms of tadpoles and aquatic adult amphibians resemble those of fresh-water teleosts. Water enters the body osmotically, but reduced skin permeability keeps this at a minimum, and it is compensated for by the production of dilute urine. Salts are retained by the kidneys but some unavoidable loss occurs. This is made up by dietary salt intake and by active absorption through the gills of tadpoles and the skin of adults.

Some amphibians, especially those of desert regions, are extremely tolerant of desiccation; they can lose tip to 50 per cent of their body water without suffering lasting damage. Such animals can generally absorb large quantities of water when the opportunity arises and, thus, make up their losses quickly. In desert forms the urinary bladder absorbs water from the urine and, interestingly, this reabsorption is also under the control of the antidiuretic hormone. Desert amphibians of the American Southwest estivate during the dry seasons and become active only during the annual summer rains when desert ponds may exist for a few days or weeks. A shortened larval period prepares the new generation for estivating before the rains cease. Estivation occurs underground where the animal can retain body water for long periods of time.

Among some amphibians, reptiles, and birds there is a reduction in the number and size of glomeruli. This is especially true of the reptiles and birds of desert areas where water retention is most critical. Reduced glomerular blood circulation also occurs in certain reptiles; it is more pronounced in lizards and snakes than in the more aquatic forms such as crocodiles and most turtles. These adaptations to a desiccating environment are reminiscent of those which have occurred in marine teleosts. The urine volume of reptiles, especially snakes and lizards, and that of birds is also reduced by the excretion of nitrogenous wastes in the form of uric acid. Uric acid is quite insoluble and can be excreted with little water loss as a semisolid paste. The excretion of such wastes in the form of soluble urea in other animals involves a much greater water loss. Birds and mammals have acquired additional control over water loss by the addition of the loop of Henle to the nephron tubules. Lower vertebrates lack this segment. The collecting tubules of birds and mammals respond to the antidiuretic hormone by increased water reabsorption as does the amphibian urinary bladder. This is an example of the older hormonal mechanism acting upon a newly evolved structure. The total length of the loop of Henle generally corresponds to the water retention requirements of the animal.

Desert rodents, such as the kangaroo rat Dipodomys and the fish-eating bat Pizonyx, both of the American Southwest and the western coast of Mexico, show this adaptation to desert conditions. The kidneys of such animals have a single renal pyramid that extends through the renal pelvis and into the ureter. This long pyramid contains greatly elongated loops of Henle and similarly elongated collecting tubules. Thus, the osmotic gradient of the renal medulla is also longer, and this enables the animal to produce an even more hypertonic urine. Both Dipodomys and Pizonyx obtain all of their water with the foods they eat; neither drinks water in the normal habitat Foods provide a fair water source for the fish-eating bat, but the diet of the kangaroo rat consists entirely of dry seeds. Experimentally it is possible to induce kangaroo rats to drink water by feeding them a high protein diet. Extra protein metabolism results in extra nitrogenous waste to be excreted through the kidneys. This, in turn, demands extra urinary water loss. Under such conditions the kangaroo rat will accept and can survive on sea water.

The mammalian kidney is a bean-shaped organ composed of a superficial cortical layer and a deeper medullary laver, both surrounding a central cavity. The whole is enclosed by a capsule of connective tissue. At the hilum, which corresponds to the "eye" of a bean, are the communicating attachments of the renal artery and vein and of the ureter leading to the urinary bladder. Within the central cavity of the kidney, the ureter expands to form a larger chamber, called the renal pelvis. The medulla of the kidney is divided into lobes, which are called pyramids because of the roughly conical shape of each. The points of the pyramids are directed inward and each is covered by a bell-shaped extension of the renal pelvis. This extension is called a calyx. The renal artery, which is a branch of the aorta, sends branches between the renal pyramids to the cortical region and, similarly, the renal vein, which drains into the inferior vena cava, has interlobar tributaries. Within the kidneys are more than a million microscopic, functional units called nephrons.

Each nephron consists of an elongate tubule closely associated at one end with a group of blood capillaries (glomerulus) and opening at the other end into a collecting tubule. The collecting tubules each receive a number of nephron tubules in this way and open, in turn, through the tips of renal pyramids into the associated calyx. The calices drain through the renal pelvis into the ureter and from there into the urinary bladder. Each nephron tubule is divisible into four morphologically distinct portions. The first of these, Bowman's capsule, is an expanded, cup-shaped region wrapped about the glomerulus. It opens into a proximal convoluted portion. A similar but distal convoluted portion opens into the collecting tubule at the other end of the nephron. Between the proximal and distal convoluted tubules is the hairpin-shaped loop of Henle. Of the structures mentioned, only the loops of Henle pass into the medulla of the kidney. The rest of the nephron lies in the cortex. This fact accounts for the faintly striated appearance of the medulla as compared with the more homogeneous appearance of the cortex when the kidney is seen with the unaided eye.

Arterial blood is supplied to each glomerulus through a terminal branch of the renal artery. This branch is called an afferent arteriole. It drains from the glomerulus through a smaller efferent arteriole and then passes through a plexus of vessels lying in close association with the tubules of the nephron. These vessels supply secondary capillary networks adjacent to the nephron walls. The glomerulus is subjected to reasonably high blood pressure through the afferent arteriole since it is almost directly supplied from the aorta with little intervening pressure drop.

The average blood pressure within the human glomerulus is about 70 millimeters of mercury. The combined membrane of the glomerular capillaries and the inner wall of Bowman's capsule is thin and permeable, and the efferent arteriole draining the glomerulus is smaller than the afferent arteriole supplying it. Consequently a high filtration pressure forces fluids through the membrane and into the cavity of Bowman's capsule. The fluid passed through the membrane is called the glomerular filtrate and is essentially the same as blood plasma except that it does not contain blood cells or large dissolved protein molecules.

All of the other normal constituents of blood are present, however. If the glomerular filtrate was allowed to flow through the nephron unchanged, all these substances would be lost to the body in the urine. Quite clearly, then, much of the content of the glomerular filtrate must be reclaimed by the body as this fluid flows through the nephron tubules. The efferent arteriole draining the glomerulus enters a peritubular plexus of vessels supplying capillaries in close association with the nephron tubules. Reabsorption of materials from the glomerular filtrate occurs through the tubule walls, and these materials reenter the blood in the peritubular capillaries. Such reabsorption occurs largely by active transport. This reabsorption is, of course, a selective one; materials such as sugars and amino acids, which the body cannot afford to lose, are reclaimed, whereas waste products such as urea are allowed to pass into the urine.

Then, glomerular filtration is a non-selective process while tubular reabsorption is highly selective. In this manner the average human renal blood flow of some 1100 liters per day yields about 180 liters of glomerular filtrate. After reabsorption from the nephrons, only about 1.5 liters remains to be passed from the body as urine. Certain substances are very thoroughly reabsorbed from the nephrons, and these substances are called high threshold substances. The term comes from the fact that high blood levels of these substances can be reabsorbed from the nephron tubules. The reabsorption capacity (threshold) of the kidney is very high for these substances. Sugars (e.g., glucose) are high threshold substances and blood glucose levels must be very high before the renal threshold is exceeded and glucose appears in the urine. Other materials are known as low threshold substances because the kidney is capable of recovering little or none of them. These are materials such as urea which are normally present in the urine. Of particular interest in physiology are substances for which the renal threshold is variable. These materials may be reabsorbed in greater or lesser quantities depending upon the conditions arid requirements of the moment By varying its threshold for such materials the kidney maintains appropriate balances within the body.

Water and certain salts, especially NaCl and KCl, are variable threshold substances. Certain other substances are actively secreted from the peritubular capillaries into the nephron lumen and thence, are eliminated with the urine. Potassium ion is treated in this way to some extent as are certain experimentally introduced molecules such as para-aminohippuric acid (PAH).

Mechanisms of locomotion

There are three distinct kinds of muscle present in the vertebrate body, namely, skeletal, cardiac, and smooth. We shall consider the skeletal muscles in greatest detail because there are more of them, because they have been more thoroughly studied, and because the other kinds differ from them in function in only a few ways. All three kinds of muscle exhibit contractility, a property exhibited by almost all cells but developed in muscle to a much greater degree. Muscle cells are all elongate cells which shorten in length when they contract. It is important to remember that the muscle cell accomplishes physical work only when it is contracting. The process of relaxation is not an active physical process. Therefore, a muscle can only pull two points closer together; it cannot push them further apart. Skeletal muscles are attached at each end, sometimes by way of an elongate tendon, in most cases to bones of the skeleton. In a few cases they may be attached to skin rather than to bone. Usually, then, the contraction of skeletal muscle pulls the attached parts of the skeleton closer together. Generally one attached end of the muscle, called its origin, remains relatively fixed in position, while the other end, the insertion, moves when the muscle contracts. For example, the biceps brachii muscle of the arm originates from the scapula bone in the shoulder and inserts on bones of the forearm just beyond the elbow.

When this muscle contracts the origin (shoulder) remains fixed but the insertion (forearm) is moved upward toward the shoulder. Since levers move in both directions, different skeletal muscles oppose each other. Such muscle pairs are called antagonists; one moves the lever in one direction and the other moves it in the opposite direction. The biceps brachii muscle is opposed by the triceps brachii muscle located on the opposite side of the humerus. The biceps brachii flexes the elbow and the triceps brachii extends it. Sometimes more than a single; muscle is involved in a given movement of a skeletal lever.

Such cooperating muscles are called-synergists. Often in groups of muscles acting generally as synergists, individual muscles actually produce minor variations of movement, depending upon which members of the group are contracting and to what degree. A series of muscles located in the forearm cooperate in various combinations to bend the wrist in any direction desired. Cardiac and smooth muscles are generally found in the walls of hollow organs. Contraction of these muscles causes constriction of the organ so that its lumen becomes smaller. Rhythmic, repeated contractions result in expelling the blood contained in the heart. In blood vessels the contraction of smooth muscle constricts the diameter of the vessel and thus raises the pressure and velocity of the blood. In the intestinal tract smooth muscle contractions result in peristalsis and related kinds of movement that mix and drive the contents of the gut forward. For these kinds of muscle there are often no antagonists since the distending force of materials inside the heart, vessels, or gut has an opposing force. Dilation occurs without the necessity of any muscular activity.

Each muscle fiber is innervated by a branch of a motor neuron by way of a neuromuscular junction near its midpoint. Each muscle fiber contains several hundred longitudinally arranged myofibrils, which are the contractile units of the muscle. Under the optical microscope each myofibril appears to be marked by prominent cross-striations. These cross-striations occur in a definite repeating sequence and have been designated by letters Of the alphabet. The dark A band, with a lighter H band through its center, alternates with a light I band, which has a darker Z band through its center. The portions between successive Z bands arc considered to be units of the myofibril and are called sarcomeres. It is this appearance of the cross-striations, of course, which has given skeletal muscle its other common name: striated muscle. The reason for the apparent cross-striations is made clear when one looks at a myofibril by means of the electron microscope. The greater resolution of this instrument permits one to see that each myofibril contains about 2500 longitudinally arranged filaments of two different kinds. There are thin filaments composed of a protein called actin and thicker filaments composed of a different protein called myosin. The cross-striations are apparent because these two kinds of filaments interdigitate and partially overlap with each other and because of the presence of interrupting walls that correspond to the Z bands. These walls, because they completely separate one sarcomere from another, are called Z discs or Z membranes. The electron microscope also reveals that the thick myosin filaments have tiny spurlike processes extending from them toward the thin actin filaments. The endoplasmic reticulum of skeletal muscle is a specialized structure known as the sarcoplasmic reticulum. It consists of two sets of tubules: (1) Fine tubules extend lengthwise parallel with the myofibrils; (2) these are connected by larger tubules that are oriented at right angles to them. The larger tubules have expansions called triads in the regions next to the interdigitation or overlap of the actin and myosin filaments.

The sarcoplasmic reticulum is believed to function in the triggering of muscle contraction following the delivery of a motor nerve impulse to the muscle. Nerve impulses result in the release of transmitter substances at synapses and neuromuscular junctions. The transmitter substance at skeletal neuromuscular junctions is acetylcholine. During contraction a great deal of energy is liberated from ATP, but most of this appears in the, form of heat. The efficiency of the muscle in converting ATP energy into actual muscular work is only about 20 to 25 per cent, depending upon the velocity of the contraction. The loss of energy is largely due to frictional resistance within the muscle cells. Interdigitation of filaments necessarily displaces the fluids of the cell matrix or sarcoplasm; this material has a fairly high viscosity. Other mechanical factors also contribute to the internal frictional resistance of muscle.

Each individual muscle fiber has a stimulus-response threshold. That is, in artificial stimulus delivered to a muscle fiber must have some minimal intensity or magnitude in order to trigger a contraction of that muscle. Stimuli of lesser magnitudes are without visible effects. If the stimulus is of sufficient magnitude to produce a contraction, that contraction occurs to a degree characteristic of the muscle fiber in question. Stimuli of greater magnitudes do not produce greater contractions. This is known as the all-or-nothing effect; the muscle fiber contracts completely, or not at all, depending upon whether its stimulus-response threshold is or is not exceeded.

Lateral fins of sharks and other fishes also act as hydrofoils, used to execute banking and turning maneuvers or as brakes to achieve sudden stops. The development of such independent movement of lateral fins required, of course, the specialization of certain myotomes in the region of the fins. The streamlining of fishes is, of course, an important factor in efficient and rapid swimming. Variations in body shape occur frequently but they are nearly always associated with a reduction in maximum swimming speed. Often protective gains are achieved by such atypical shapes, however, and thesc have survival value as well. The bilateral flattening of fishes, for example, is quite common.

Angel fish are bilaterally flattened to such an extent that they resemble the plants growing in their native South American rivers. In other cases extreme flattening makes the animal more difficult for predators to detect when seen edge on. Typically, such fish are bottom feeders and are not called upon to swim rapidly. Rays and skates are dorsoventrally flattened and are also bottom feeders. Rays are so highly modified that typical swimming movements are no longer possible; instead, successive waves of the lateral body margins propel the fish forward. This movement still involves the same successive myotomes but the motion is up and down rather than side-to-side as in typical fishes. The body shape of the sea horse has been so completely modified that myotomes are no longer used for swimming at all. Instead, these very poor swimmers stand vertically in the water and propel themselves by wavelike motions of the dorsal fin.

The elongate bodies of eels reduce swimming efficiency but have the advantage of allowing snakelike locomotion on land. Some fishes can swim so rapidly that they are able to leap for considerably heights out of the water. Salmon and tarpon, in particular, are capable of such leaping. Flying fishes prolong the leap by using greatly enlarged lateral fins as gliding wings, and the flying gurnards actually flap the fins while in flight. The lateral fins of certain bottom dwellers are modified as spinelike levers and are moved as legs. Such fins are of little use as stabilizers or steering mechanisms during swimming but they permit easy locomotion across the bottom. Of course, the use of such fins has necessitated the development of stronger muscles to move them, and in some cases such fishes are also able to walk across land.

The structure of electrogenic organs is generally quite similar in all fishes in which they occur. Each consists of a series of thin modified muscle cells called electroplaxes arranged in columns like stacks of coins. Each cell is innervated on one side only. At rest, each electroplax carries a membrane change much like that of any other cell-negative on the inside with respect to the outside. Each column of electroplaxes, then, consists of alternating membranes of opposite polarities. The first membrane is charged positively above and negatively below; the next membrane is charged negatively above and positively below, and so forth When stimulated by way of the nerve on one side of the electroplax the polarity of that side is momentarily reversed, as is that of a nerve membrane. At that moment, then, the column of electroplaxes consists of a series of membranes which are all charged in the same direction. Every membrane is then negative above and positive below. In this configuration the membranes of the column are like multiple plates in a battery that are wired in series. A salamander has an elongate, somewhat streamlined body much like that of a fish. Ordinarily it moves by means of its limbs but when frightened it often allows the legs to trail unused and travels on its belly by means of typical swimming movements. Even when walking on its limbs, successive bending movements of the body occur, and the movement of the limbs is coordinated with this. This pattern of leg movement, arising as it did from swimming activity, remains the pattern in all four-legged vertebrates.

Even adult humans, who do not use the "forelimbs" in walking, still swing them at the sides in the same motion pattern. Long spinal reflexes were established early in vertebrate evolution and continue to be functional. The development of legs that were more than stiff levers like those of certain fishes required the further modification of the associated muscle groups. Walking on legs places new stresses on the vertebral column, which must now serve as an arch to support the body weight.

Locomotion of lizards is much like that of salamanders. The body successively curves from side to side and the limb movements are coordinated with it. In a few cases lizards can run on the hind limbs but only when frightened. Snakes have even more primitive locomotion. It is essentially "swimming on land," as an eel does. A few snakes move by dorsoventral undulations like those of a caterpillar, but much less noticeable. The sidewinder rattlesnake Crotalus cerasles employs a different variation, one which consists of a series of looping motions during which only two portions of the snake contact the ground at any one time. This corkscrew-like motion leaves a trail of long parallel marks in the sands of the desert where this snake occurs. Like some modern lizards, many extinct reptiles were able to move about on the hind limbs only. When forelimbs are relieved of any important role in walking, they become available for other tasks, and the modification of our own arms and hands reflects this fact.

Bipedal walking preceded the development of wings in the first birds since, as it is believed, birds evolved from reptiles such as these. Two theories have been offered to explain how the evolution of bird flight may have come about one theory suggests that a group of bipedal reptiles developed the habit of flapping the forelimbs up and down when running. The least bit of lift attained by such flapping would have been advantageous since it would have permitted longer and longer strides. Presumably, then, these strides would have become gliding leaps and finally sustained flight as other adaptive changes took place in the body to permit it. The second theory of flight is probably better supported by available evidence. This suggests that the ancestral reptile-bird was a tree-dweller that developed forelimb adaptations as an aid in leaping from tree to tree. Progressive development extended these leaps until sustained flight became possible.

Archaeopteryx is the only intermediate fossil species known, and the reptile-bird nature of this animal is obvious. Bird locomotion differs markedly from that of most other vertebrates in that the forelimbs are moved together in flight instead of alternately as in bipedal motion. Even on the ground many birds hop rather than walk. Most flightless birds or birds that fly very weakly, however, walk or run in the typical vertebrate fashion. When a flat plane is moved through the air, the medium is cleaved by it, thus resulting in an equal pressure on both sides of the plane. This parting of the medium requires energy, of course, and the reuniting of the medium behind the trailing edge of the plane creates friction or drag. If the shape of the plane is modified so that its surfaces coincide with the natural pathways of the parted and rejoining air, the changes in pressure on the two sides and the drag are minimized so that less energy is required to move the plane forward. If the tipper and lower surfaces of the plane are not alike, the pressure developed on the two sides will also be different. If, for example, the plane is flat on the underside but streamlined above, the air will be thrown further aside below and thus create more pressure there. This is a typical airfoil; when it is moved forward through the air, the increased pressure below lifts the plane. If the leading edge of the plane is tipped upwards, the lift is still further increased but so is the drag, especially at higher speeds.

In any case, at some particular minimum forward speed, the drag becomes maximal and the lift becomes minimal. At this point the plane stalls and falls. A wing slot in the leading edge of a wing can be so constructed that it directs a layer of air over the wing to reduce drag and, therefore, lower the stalling speed. Birds make use of the airfoil shape of wings and even have leading edge slots. A small structure called the alula can be moved forward to create such it slot. Unlike aircraft, birds are also able to change the shape and angle of the airfoil to some extent while in flight and thus vary the lift and drag qualities to meet their immediate needs. When landing, aircraft commonly slow the forward speed by lowering flaps from the trailing edge of the wing. Birds accomplish the same thing by lowering feathers of the trailing edge of the wing as well as feathers of the tail.

It is commonly supposed that the downstroke of a bird's wing is for the purpose of lifting the bird upward; this, however, is not true. Actually the bird obtains lift in the same way that an aircraft does, namely, from the cross-sectional shape of the wings and the forward motion of them through the air. The flapping of the wings is for the purpose of maintaining forward motion.

In other words, it is really "swimming" through the air. On the downstroke there is also a strong forward thrust imparted, and in some birds the upstroke also drives the bird forward because of a change in the angle of the wings. In most cases, however, the upstroke is merely a recovery stroke. The wing is folded close to the body and the feathers separate like the strips of a venetian blind so that little air resistance is encountered during the upstroke.

Co-ordinating mechanisms

Response of a receptor to an environmental stimulus is merely an extension and specialization of a property common to all living tissue, namely that of irritability. The specificity of receptor response to a particular kind of stimulus results from the individual characteristics and physical, locations of receptor cells and from the modifying influences of surrounding cells.

In some cases such surrounding cells form parts of the total receptor organ. A single example will serve to illustrate the point. The receptor cells of the vertebrate eye are most sensitive to and respond most effectively to changes in light intensity. This is partly due to the intricate array of surrounding tissues that are specialized to form a cameralike organ with automatic exposure control and focusing mechanisms. It is also due to the fact that the receptor cells contain photosensitive chemical substances.

These same cells will also respond to other kinds of environmental stimuli. Pressure exerted with a fingertip against the side of the eyeball is an effective stimulus, but because the brain can interpret signals from the retina of the eye only in terms of light, the conscious sensation is of that type. Electrical and thermal stimuli elicit similar responses. Thus, retinal receptors retain the general irritability of all living cells but have heightened sensitivity to photic stimuli and are situated where other kinds of stimuli are unlikely to reach them. The response of an animal to changing environmental conditions is generally complex, requiring the integration of several kinds of information from different receptors and involving, perhaps, more than a single kind of physiological adjustment. Often information stored as memory—a special input permitting the animal to learn and profit from experience—is also utilized. Such complex integration requires the presence of a computerlike central nervous system equipped with data storage or memory banks. The central nervous system must quickly integrate all of the input available to it and provide an appropriate output which, through initiating activity of effector organs, results in purposeful physiological responses. These responses and their results must be monitored by the same or other receptors, or both, and the "feedback" input must be integrated in the same way. The monitoring process permits the responses to be continued, modified, or stopped as required to achieve a desirable final result.

An extremely mechanistic view of the vertebrate animal, then, would show it as a complex, highly automated machine that combines sensitive receptors of many kinds, a finely controlled set of effector organs, and an extremely efficient central computer. These three basic units are in constant communication through networks of peripheral nerves so that all activity can be coordinated and purposeful. Nerve impulses the signals that pass along nerves-are of only one kind, or in the activity of the brain destined to cause the contraction of skeletal muscle. The differences in the kind and degree of effect depend upon (1) the origin and destination of the impulses, (2) the number of similar conducting nerve cells (neurons), and (3) the number of impulses per unit length of-time traveling along each neuron. The number of transmitting neurons and the frequency of impulses transmitted by each are factors related to the intensity of the original stimulus. The receptor "translates" actual environmental changes into coded patterns of nerve impulses. In electrical recording instruments, such receptorlike parts are called transducers; these convert specific kinds of environmental events into recordable electrical signals in an analogous way.

Within the central nervous system, the incoming coded signals may require integration with other sensory input signals and perhaps with stored memory data. The central nervous system initiates patterns of nerve impulses which travel through other peripheral nerves to various effector organs. These nerve impulses are identical in character with those of sensory nerves, but because they are transmitted to muscles, glands, and other effectors, they result in adjustments of physical position or of physiological processes. As in sensory coding, the frequency of impulse transmission and the total number of parallel neurons involved determine the degree of effector response.

Each effector organ "decodes" the signals and responds in the proper way and to the appropriate degree. In very simple, somewhat stereotyped stimulus-response situations no great amount of central integration is necessary because there is only a single kind of appropriate response. Generally that response must occur immediately without any delay imposed by the central nervous system. Placing a finger in a flame, for example, results in the immediate withdrawal of the finger-the only reasonable response to such a painful stimulus. The withdrawal of the finger involves the activation of the flexor muscles of the arm and forearm, and the response is called, therefore, a flexor reflex. It actually takes place before the brain has consciously interpreted the sensory input as pain. This is true because the reflex mechanism takes place through a simple and direct nervous are involving only a portion of the spinal cord. An all-but-direct link between the sensory and motor neurons within the spinal cord results in an immediate contraction of flexor muscles, causing withdrawal of the burned finger.

The more intense the stimulus, the more violent is the withdrawal by reason of the involvement of greater numbers of neurons in both the sensory and motor portions of the reflex arc. In fact, if the stimulus is intense enough, the response may radiate to other portions of the body as well so that in addition to removing the finger, one may even leap backwards away from danger almost before realizing consciously that an accident has occurred. Whenever any muscle is at rest, it is nevertheless partly in a contracted state. This state of partial contraction is known as tonus.

Now, when a given muscle contracts, either reflexly or voluntarily, the tonus of opposing muscles must be reflexly inhibited so that they do not unnecessarily oppose the action of the contracting muscles. For example, contraction of the biceps muscle, which flexes the upper extremity at the elbow, would be opposed by the normal tonus of the triceps muscle, which extends the same limb. Whenever the biceps is stimulated to contract, the tonus of the triceps is inhibited, and this occurs by way of spinal cord level reflex connections.

The withdrawal of the finger from the flame occurs faster and more efficiently because of this inhibition of tonus in the antagonistic triceps muscle. Such simultaneous and opposite controls of antagonistic muscles involves the principle of reciprocal innervation and occurs reflexly in the spinal cord. Completely unopposed muscle responses, especially when they occur suddenly and strongly as they would in a flexor reflex induced by a painful stimulus, tend to be uncontrolled and jerky. Two other spinal cord reflex mechanisms act to smooth such actions by first opposing the response slightly and then releasing it as follows: (1) The contraction of any skeletal muscle results in the stretching of an antagonistic muscle. Contraction of the biceps necessarily results in stretching the triceps muscle. Sensory receptors within the stretched muscle are stimulated and act by way of a simple, cord level reflex arc, to cause a contraction of that muscle sufficient to resist the stretch. This is called a myotatic or stretch reflex and it acts to damp the sudden onset of any skeletal muscle activity. A sudden contraction of the biceps is, thus, momentarily resisted by contraction of the opposing triceps muscle. (2) If the initial contraction of the biceps continues, the increasing stretch of the triceps muscle then activates still other receptors within the triceps tendons, and these, also by way of a simple reflex arc, initiate a reverse myotatic reflex.

This reflex results in a sudden disappearance of the myotatic response; the further contraction of the biceps is then unopposed. Since the pull of gravity tends to cause limb flexion and, therefore, the stretching of extensor muscles, certain myotatic reflexes function also as antigravity responses.

Suppose, for example, that flexion of the lower extremity begins to occur at the knee of a standing human. The force of gravity will tend to continue such flexion and, if totally unopposed, will cause the individual to fall to the ground. Flexion at the knee, however, excites the stretch receptors of the extensor muscles of the limb, and this elicits a myotatic reflex. The extensor muscles contract and resist the further flexion of the knee joint. This may stop or even reverse the flexion (extend the limb) to prevent falling, or if the force causing the flexion is too great (perhaps the individual desires to kneel or fall or the load he is bearing is too great), the reverse myotatic reflex soon takes over and continued flexion occurs unopposed. This sudden unopposed flexion is sometimes spoken of as the clasp-knife effect. The antigravity myotatic reflex just discussed can he demonstrated by tapping the patellar tendon just below the kneecap to produce the familiar knee jerk or kicking (extension) response.

Similar antigravity responses can be demonstrated at the ankle joint, and even though humans do not walk on all four limbs, antigravity responses for the elbow and wrist joints are present. Interestingly, the South American tree sloth which hangs upside down from tree limbs, uses flexor muscles rather than extensor muscles to resist the pull of gravity. In this-animal typical myotatic responses involve the contraction of flexor muscles. The pressure of the soles of the feet against the ground stimulates receptors in the skin of the soles.

By way of simple reflex arcs this input into the spinal cord results in extension of the legs. This supportive reflex can easily be demonstrated by pressing lightly against the footpad of a reclining dog or cat. Even in this position the animal will respond by extending the stimulated limb. If the limb of the animal is first flexed to a considerable degree, the effectiveness of this stimulus disappears. This reflex is of importance in walking since the limb extends against the ground while in contact with it. Flexion of the limb removes the contact and, therefore, the tendency to extend the limb.

Further flexion then becomes easier. However, there are simple spinal cord reflexes that are set in motion by a stimulating event on one side of the body and result in muscle activity on the opposite side of the body. These are called contralateral reflexes. One such contralateral response results from the crossed extensor reflex. When the right leg is flexed, the crossed extensor reflex causes an immediate extension of the left leg. The value of such a reflex when maintaining an upright position while walking or while shifting the weight of the body from one foot to the other is quite obvious. Most spinal cord reflexes tend to fatigue and disappear rather quickly after they are evoked. The same reflex is then more difficult to elicit a second time until recovery has taken place. At the same time opposing reflexes are easier to elicit This phenomenon is known as the *rebound phenomenon.*

In quadruped vertebrates the alternate activity of the hind legs is coordinated with that of the forelegs to give the animal a characteristic four-legged pace. This coordination is accomplished by long spinal reflexes which are similar in principle but slightly more complex in anatomical linkage than the reflexes discussed previously. Some animals have more than a single pattern of four-legged coordination. Movements of the head and neck are also coordinated with body and leg movements by special neck reflexes. These also operate on a spinal cord level and are little more complex than the long spinal reflexes just discussed. Neck reflexes keep the head steady despite the side-to-side movements of the body during walking. They also direct the head and limbs toward common goals in leaping and other special activities.

As a quadruped vertebrate walks in a straight line, the shoulders may move from side to side. If equal and opposite bending of the neck did not also occur, the head would be swung from side to side as well. This characteristic movement of the shoulders and of the head is a consequence of the basic pattern of locomotion common to all vertebrates.

Even in humans, in which no lateral movements of the shoulders ordinarily accompany locomotion, one sometimes sees rhythmic side movements of the head coordinated with walking. If the head of a standing cat is lifted by placing a finger beneath its chin and raising it so that the animal is looking upward, the hind legs will flex and the fore legs will extend. This places the animal in a crouching position appropriate to leaping upward in the direction of its gaze. Conversely, if the animal's head is pressed downward, the forelegs will flex and the hind legs will extend. This again places the animal in the appropriate position for leaping downard in the direction of its gaze.

Similarly, turning the cat's head toward the right will cause flexion of the right foreleg and extension of the left foreleg- movements appropriate to stepping toward the right All of these responses are mediated by simple neck reflex connections within the spinal cord and, therefore, require no integration within-the brain. The conditioned reflex is well known to most people; it was demonstrated by Pavlov that ringing a bell and feeding a dog at the same time could establish such a pattern. After a time the bell alone becomes an adequate stimulus to elicit salivation in the dog by way of reflex pathways. It is really impossible to draw a sharp line between this simple kind of conditioned reflex and a variety of more complex behavior.

The complete nature of the nervous impulse is not entirely known, but it involves both electrical changes at the surface of the neuron and internal changes, which proceed in a wave-like manner at speeds up to 100 meters per second in some mammalian nerves. Nerve cell membranes, like the membranes of other living cells, have an electrical charge when they are at rest. When a neuron is stimulated, the permeability of the membrane at the point of stimulus is altered; as a result, sodium ions and potassium ions are able to diffuse readily through the membrane in accordance with the existing concentration gradients for each. That is, sodium ions diffuse inward and potassium ions diffuse outward. The permeability becomes greatest for sodium ions, and they rush into the cell rapidly. Potassium ions leave the cell somewhat more slowly, since the permeability to these is not increased as much. The result is that momentarily there is a large combined population of sodium and potassium inside the cell a population sufficient to counterbalance the negatively charged organic molecules there and give the inside a net positive charge. At the same time, the outside surface of the membrane has been depleted of sodium ions, and it momentarily has a negative charge with respect to the inside.

Thus, the change in membrane permeability produces a momentary reversal of membrane change; the membrane becomes negative on the outside and positive on the inside. The continued outward diffusion of potassium ions removes positive charges from the inside and restores them to the outside. This continues until the distribution of both sodium and potassium has reached an equilibrium, and, as a result, the membrane potential begins to return toward the resting state. At the same time, however, the "sodium pump" of the membrane increases its activity and the permeability characteristics of the membrane return to normal. As a result, sodium ions are again pumped outward, potassium ions are accumulated inside, and the resting membrane potential is reestablished. The momentary reversal of membrane charge and the subsequent reestablishment of the resting potential just described are together called an action potential.

The action potential coincides with and is at least a part of the nervous impulse. It sweeps over the surface of the nerve cell from the point of origin to all other parts of the cell in a wavelike fashion. As one portion of the neuron membrane is reversed in polarity by the action potential, this is in itself an adequate stimulus to adjacent portions of the membrane so that they follow suit. The action potential, therefore, spreads along the membrane with recovery of the resting membrane potential following behind. This conductive excitatory state, as it is sometimes called, appears as a wave of negativity of the outer surface of the membrane.

Certain large, rapidly conducting neurons are covered with layers of a substance called myelin; smaller, slower-conducting neurons have no such covering. Myelin is a lipid material that provides electrical insulation and protection for the neuron. The myelin is associated with Schwann cell membranes, which are wrapped in a spiral fashion around the nerve cell process (axon). This covering is interrupted at intervals by nodes of Ranvier, at which points the neuron membrane is exposed. Ions cannot flow easily through the insulating myelin but pass through the neuron membrane with ease at the nodes of Ranvier. As a result, the action potential in such neurons appears to leap from node to node.

The electrical disturbance spreads through the insulated internodes by electrical conduction, and it is only at the nodes where membrane polarity is momentarily reversed. This minimizes the task of the "sodium pump" and speeds the entire process of action potential transmission so that myelinated neurons often carry impulses at up to 500 times the speed achieved in nonmyelinated neurons. The fastest neurons are used for reflex mechanisms involving skeletal muscles, whereas slower neurons act in connection with smooth muscles and in other ways. The application of a stimulus to a sensory receptor induces in it a receptor potential or, generator potential. This potential constitutes an effective stimulus to the associated sensory neuron and elicits an action potential in it. Often the receptor potential persists for a time so that as soon as the associated neuron has reestablished its resting potential, it is restimulated. In this way a single stimulus can evoke trains of nerve impulses traveling toward the central nervous system. The numbers of such impulses and the frequency at which they occur in a sensory nerve depend upon the magnitude and the duration of the receptor potential, and this, in turn, is dependent upon the intensity of the stimulus.

Human brain

The development of the cerebral hemispheres plus the repositioning of parts to accommodate the upright posture of man somewhat distorts the pattern, however. The cerebral hemispheres hide from view the other portions of the forebrain, most of the midbrain, and a part of the hindbrain when the entire brain is seen in lateral view. The cerebrum contains two large cavities (ventricles I and II) that the lateral to the midline and a third cavity (ventricle III) in the midline.

Much of the substance surrounding these cavities is composed of nerve tracts and collections of nerve cell bodies; the latter the in regions called ganglia or nuclei. The exposed surface of the cerebrum consists of nerve cell bodies arranged to form the cortex of the cerebrum. This surface cortex is marked by a number of deep fissures and shallow sulci that give it a convoluted appearance. The convolutions between sulci (singular sulcus) and fissures are called gyri (singular gyrus). At first glance these appear to be distributed in a random pattern. Actually the same pattern of fissures, sulci, and gyri occurs on every human brain, and each of these structures is given a definite name by anatomists. We will be concerned here with only a few names of landmarks on the surface of the cerebrum. The right and left cerebral hemispheres are separated from each other by a deep sagittal fissure.

Each hemisphere is further divided by a vertical central fissure and a diagonal lateral fissure. These, together with other more arbitrary landmarks, divide the cerebral cortex into four immediately visible lobes, namely the frontal lobe anteriorly, the temporal lobe laterally, and the parietal and occipital lobes posteriorly. The cavities within the cerebral hemispheres (ventricles I and II) extend forward into the frontal lobes, backward through the parietal lobes and into the occipital lobes. They also have extensions laterally and anteriorly into the temporal lobes. Each of these lateral ventricles communicates through a small interventricular foramen (hole) with the single median cavity (ventricle III). This third ventricle lies between and slightly below the level of the lateral cavities. From the third ventricle a narrow canal, the cerebral aqueduct; passes through the midbrain region to open into a fourth cavity (ventricle IV) located in the hindbrain. The fourth ventricle, in turn, communicates with the tiny central canal of the spinal cord and also opens to the surface of the brain through three small openings, namely the foramina of Luschka and Magendie. The outer surface of the entire central nervous system is covered by a thin, closely adherent membrane called the pia mater. This membrane dips into all of the fissures and sulci following faithfully the contours of the central nervous system.

Over the pia mater is a second membrane called the arachnoid membrane. It is attached the pia mater by delicate spider-web-like strands of tissue; the space between the arachnoid membrane and the pia mater is crossed only by these strands. This is the subarachnoid space, and like the internal ventricles and passages, it is filled with cerebrospinal fluid. The arachnoid membrane dips into the fissures and the larger sulci but does not follow as closely the contours of the central nervous system as does the pia mater. Thus, the subarachnoid space is relatively large in certain places. Finally, a third membrane, the dura mater, encloses the whole.

The dura mater is a tough, thick layer which is doubled in some places. In such places the inner layer dips into the major fissures of the brain while the outer layer follows more closely the inner surface of the skull. The dura mater is, indeed, continuous with the connective tissue covering (lining) of the bones surrounding the central nervous system. This connective tissue, which covers all bones, is called the periosteum. In regions where the dura mater is doubled there are relatively large venous blood sinuses between the two dura layers that serve as the major routes of venous drainage from the brain.

Like the two other layers of meninges described above the pia mater and the arachnoid membrane the dura mater also extends downward to envelop the spinal cord. In this region it even extends laterally to enclose the first portion of each spinal nerve. Branches of cerebral arteries are closely associated, with the walls of all four of the internal ventricles of the brain. The endothelium of associated capillaries together with the lining membranes of the ventricles (ependyma) form a specialized membrane in the part of each ventricle in which the cerebrospinal fluid, a special filtrate of blood plasma, is formed. This specialized membrane is called the choroid plexus, and the cerebrospinal fluid formed by it is continuously secreted into the cavities of the ventricles.

As the fluid is produced it flows through the communicating passages between the ventricles and escapes through the openings from the fourth ventricle (foramina of Luschka and Magendie) into the subarachnoid spaces surrounding the central nervous system. In the regions of the venous blood sinuses the arachnoid membrane here and there penetrates through the inner layer of dura mater and into the blood sinuses. At these points, called arachnoid granulations, cerebrospinal fluid passes from the subarachnoid spaces into the venous blood and is carried away with the venous drainage.

Thus, cerebrospinal fluid, derived from arterial blood through the choroid plexus, circulates through the interior of the central nervous system, bathes the exterior of the central nervous system, and is ultimately returned to the blood through the arachnoid granulations. This fluid is like blood plasma in many respects but it also has some important differences, i.e., it lacks blood cells and its protein content is quite different. Since it also lacks the buffer systems that are present in blood, its pH is altered by changes in CO_2 concentration. This is probably an important factor in the controls of respiration, of blood pressure, and perhaps in other ways as well. The special permeability characteristics of the choroid plexus have given it the designation of "blood-brain barrier." It is important, for example, to determine whether or not drugs of various kinds will cross the "barrier" and thus be effective in the treatment of certain central nervous system diseases.

Within the cerebrum are many nerve tracts leading to and from lower levels of the central nervous system and interconnecting various parts of the cerebrum itself. Besides these tracts there are also several internal concentrations of nerve cell bodies including the thalami, the caudate nuclei, the lentiform nuclei, the hypothalami, and others. From the floor of the forebrain extends the stalklike infundibulum leading to the pituitary gland or hypophysis. Also prominently visible on the floor are the olfactory bulbs and tracts, the optic chiasma (where the right and left optic nerves meet to exchange fibers), the small round mammillary bodies, and the other pairs of cranial nerves.

The cerebrum and other portions of the forebrain are connected through the midbrain, or mesencephalon, to the hindbrain. The mesencephalon contains, of course, many nerve tracts connecting higher and lower portions of the central nervous system but it also contains several important nuclei that have special functions. Dorsal swellings on the surface of the mesencephalon, the corpora quadrigemina, contain nuclei that function in the integration of reflexes of body balance and optical orientation. Other nuclei within the mesencephalon serve other fundamental integrative purposes. The medulla oblongata, like the midbrain, contains nerve tracts leading to higher and lower portions of the central nervous system as well as concentrations of nuclei that serve- very fundamental nervous reflexes. The centers of control for respiration and blood pressure, for example, are found in the medulla oblongata. Other nerve tracts pass horizontally through and around the medulla linking together the right and left hemispheres of the cerebellum and forming the prominent anterior bulge called the pons cerebelli.

The cerebellum is the other part of the hindbrain and extends as paired hemispheres from the dorsal surface of the medulla oblongata. It is marked by long, narrow surface gyri separated from each other by deep sulci or fissures. These sulci branch below the surface, giving the cerebellum a very large surface area. Through experimentation and by the careful examination of the brains of persons with known functional losses, it has been possible to map the functional areas of the cerebral cortex.

Parts of the cerebral cortex function in ways that are difficult to quantify because they involve psychic factors that are not easily defined. Such characteristics as introspection, foresight, moral inhibition, ambition, conscious pleasure, and so on are relegated to these so-called psychic areas of the cerebral cortex. The cortex of part of the frontal lobes, most of the temporal lobes, and much of that within the sagittal fissure is psychic in function. The occipital cortex is the region of visual reception and interpretation. A central portion of it receives, initially, nerve impulses from the retinas of the eyes, and a surrounding portion interprets this pattern of impulses in terms of visual experience and learning. It is in this surrounding area, which we may call the visual association area, that conscious impressions of vision occur.

A similar primary auditory receiving area, surrounded by an auditory association area for conscious interpretation of sound, occurs on the superior portion of the temporal lobe. This area is not immediately visible in a lateral view of the brain, however, since it lies within the lateral fissure. The primary cortex for the receipt of general cutaneous sensations including touch, temperature, and so forth, occupies the postcentral gyrus—the gyrus which is immediately parallel to the central fissure and is a part of the parietal lobe of the cerebrum.

Posterior to this primary area and covering much of the anterior and superior portions of the parietal lobe is the conscious association area for these same sensations. The posterior and superior portions of the frontal lobe of the cerebrum contain the cortical area concerned with the conscious production of muscular activity. These are arranged in much the same way as the sensory areas just described, although these are motor in function. It is the cortex of the precentral gyrus the gyrus immediately anterior and parallel to the central fissure-that initiates the nerve impulses causing voluntary muscle contractions. The remaining motor association area of the frontal lobe cortex contains the learned patterns of activity, i.e., the memory of how to carry out various muscle activities.

Virtually all of the conscious senses, as well as the voluntary motor regions of the cerebral cortex of humans, are potentially involved in language symbolism. It is interesting, therefore, to note that a central area of the parietal lobe cortex is apparently involved in this special interpretive function. This symbolic language association area, because of its central position, lies immediately adjacent to the visual, auditory, and cutaneous sensory association areas. Moreover, there is a special nerve tract linking it with the frontal cortex areas, which are concerned with movements of the vocal muscles and those of the hands, i.e., the muscles used in producing language symbols for others to see or hear.

Each of the functional areas of the human cerebral cortex has a more or less definite internal organization in which body parts are arranged in sequence. For example, both the primary motor area (precentral gyrus) and the primary cutaneous sensory area (postcentral gyrus) contain sequential regions connected with the various body parts in almost exactly the same order as these body parts occur. This is not really surprising when one considers that nerve tracts remain in laminar organization as they ascend or descend through the central nervous system.

Body portions which are heavily innervated with either cutaneous sensory or skeletal muscle motor nerves are represented by correspondingly larger cortical regions in these primary areas. In a similar way the various portions of the retinas of the eyes are represented by corresponding portions of the occipital cortex, and the various portions of the organ of hearing of the ears are represented by sequential portions of the auditory cortex. The internal nuclei or ganglia of the cerebrum are functional in the control of gross, stereotyped motor activities in man.

In lower vertebrates, in which the cerebral cortex is less well developed, the thalami apparently have greater importance. In these animals the thalami are nearly the highest brain center and they control all motor activity; they probably also interpret most sensory input as well. In humans sensory input passes through the thalami on the way to the cortex and, similarly, cortical motor output is relayed through the thalami on its way to lower levels of the central nervous system. Much remains to be learned about the details of function of these internal accumulations of nerve cell bodies in the human cerebrum.

Beneath the cerebral ganglia and nuclei just discussed is another complex group of nuclei collectively called the hypothalamus. This contains centers that control the activities of the pituitary gland and a variety of other fundamental processes. It is apparent that sleep and wakefulness, thermoregulation, hunger, and perhaps other states are controlled by these subconscious centers in the hypothalamus. The role of the cerebellum is an extremely interesting one. It will be recalled that the precentral gyrus of the cerebral cortex initiates nerve impulses that produce voluntary movements of skeletal muscles.

A pattern of impulses originates there and is conducted into the spinal cord and out through various spinal nerves to appropriate muscles of the body. This pattern of impulses arising in the cerebral cortex, however, merely produces a rather gross muscular activity, and it is necessary to modify the pattern as the activity progresses in order that it be smooth and effective. The cerebellum monitors the motor output of the cerebral cortex and also monitors its effects. It receives feedback from the various sensory receptors within the muscles, their tendons and the joints of the skeleton so that it is constantly informed of the progress of the activity induced there. From this information it adjusts and modifies the motor output so that the activity is carried out smoothly and purposefully.

Receptor organs

Receptor organs are commonly classified in one of two different ways:

i) Exteroceptors are located on or near the surface of the body and they monitor conditions in the external environment. These include receptors for -vision, olfaction, hearing, taste, touch, and temperature. Proprioceptors are located within muscles, tendons, and joints and respond to stretching, pressure, and the like. They monitor conditions relating to body posture and motion—the kinesthetic senses. Interoceptors are located within organs and tissues other than skeletal muscle and they monitor various modalities of the internal environment. These include the pressure and chemoreceptors of the carotid sinus, the stretch receptors of the lungs, pain receptors of the intestinal tract, and many others.

ii) Receptors may also be classified on the basis of the type of stimulus modality that they are especially adapted to monitor. Mechanoreceptors respond to physical forces such as pressure, vibration, and inertia. These include receptors for touch, hearing, equilibration, and many others. Chemoreceptors respond to chemical stances and include olfaction and taste as well as the response to carbon dioxide and oxygen concentrations in the blood. Photoreceptors may be classed separately or as a special case of chemoreception, since photo-receptors actually respond to a photochemical reaction set in motion by light rather than to the light itself.

Temperature reception

Animals can be described as being cold-blooded or warm-blooded, depending upon whether they generally feel cool or warm to the touch. Cold-blooded vertebrates, are the fishes, amphibians, and reptiles; the warm-blooded vertebrates are the birds and mammals. More precisely, we can describe animals as being ectothermic or endothermic. Ectothermic animals are primarily dependent upon environmental temperatures to determine their own body temperatures. All animals produce metabolic heat, but the ectotherms are unable to vary heat production or to control heat loss by physiological adjustments. They can only maintain tolerable body temperatures by behavioral means, i.e., they must find warmer or cooler portions of their habitats as required. Endotherms, on the other hand, are able to thermoregulate by both behavioral and physiological means.

Endotherms maintain body temperatures that are higher than those of ectotherms. There is still another set of terms used to describe the temperature relationships of vertebrate groups. These terms are poikilothermic, homeothermic, and heterothermic. A poikilotherm is an animal in which the body temperature is always very close to the environmental temperature. The term is, therefore, essentially a synonym for ectotherm. A homeotherm is an animal which maintains a constant or nearly constant body temperature despite wide fluctuations of environmental temperature. Such animals are, of course, endotherms, but not all endotherms are homeothermic. Some endotherms, instead, tolerate fairly wide fluctuations of body temperature and resist further change in body temperature only when it approaches the critical limits compatible with life. Such animals are called heterotherms. Under experimental conditions many vertebrates can adjust their sensitivities to temperature extremes.

Fishes have an internal temperature that is slightly higher than that of the surrounding water but the usual difference is slight. The rate of metabolism is slow and the thermal conductivity of both tissues and water is high. Heat loss, therefore, is almost as rapid as heat production. Increased activity of the fish produces more heat, but since it then also requires more ventilation of the gills, the rate of heat loss is also increased. Temperature regulation in fishes, depends almost entirely on the selection of environmental regions which have appropriate temperatures. When species are trapped in especially warm or cold waters, they are acclimatized in various ways. Some are even able to tolerate sudden temperature changes of considerable extent.

During rainy season, a storm may increase water volume by several fold and lower the water temperature by 10^0 C or more within a matter of minutes. It is interesting to note also that during the dry season the pond may actually have minerals precipitated on the surface; the sudden rain storm, then, produces great dilution of the pond's mineral content. Thus, these fishes are also exposed to sudden and drastic changes in the salinity of their environment. The physiological means for adaptation to such salinity changes have not been thoroughly studied.

Amphibians are largely limited to behavioral thermoregulation. In some cases they withstand prolonged low temperatures by hibernation. For many amphibians thermal regulation to high temperature is quite effective by reason of their yet skin and consequent water evaporation, but the resulting loss of body water is a limiting factor. Desert amphibians estivate, a process somewhat like hibernation, by burying themselves in the soil during the hot, dry part of, the summer, while semiaquatic amphibians thermoregulate well in warm temperatures thermoregulation to low temperatures is more difficult. The skin is primarily a respiratory surface and, therefore, loses heat easily. Because the skins of reptiles are dry, these animals lose heat less easily and also have better control over water loss than amphibians. Improved renal functions also contribute to water conservation.

Behavioral adaptations to temperature fluctuations can be more sophisticated than in the Amphibia. Depending on their body temperature at the time, reptiles may select warm or cool surfaces to lie upon. During the cool desert nights, reptiles often lie upon exposed rock or paved highways to absorb the remaining, heat from the sun. During the hot day they often bury themselves in sand or crawl under other insulating materials. Reptiles may also position themselves in sun or shade so that a greater or lesser area of skin is exposed to solar radiation.

In certain birds and mammals of arctic regions the temperature of the extremities may fall well below body temperature and still receive an adequate blood supply. In such animals the tissues of the extremities are more tolerant of cold than are other body tissues —an example of regional acclimatization. The temperature differential is maintained without undue heat loss from the extremities by means of a countercurrent heat exchange mechanism. This operates on principles similar to those encountered in gill respiration and swim bladder inflation. Articles carrying warm blood into the extremities pass close to and parallel with veins carrying cool blood toward the heart. In fact, the veins often form sleeve like networks about the arteries. Much of the heat of the arterial blood is used to warm the venous blood and as a result little body hear actually enters the extremities where it could be easily lost to the cold air.

Whales and seals have such vascular arrangements in the flippers and flukes because these parts, unlike the body, are not protected by insulating fat layers. Penguins have similar vessels in the axillary regions so that heat loss from the wings is reduced. Countercurrent heat exchangers of this kind are not limited to animals inhabiting, cold climates.

Certain tropical mammals such as the sloth, the loris, and the more primitive armadillos and anteaters are extremely sensitive to environmental temperatures that are only a little below- body temperature. These animals are capable of little thermoregulation by other means, but they have vascular arrangements similar to those discussed in the preceding paragraph. Some countercurrent heat exchange occurs in the extremities of humans as well. The continued operation of a heat exchanger during the warm summer months would interfere with necessary heat loss for some animals, especially those which retain a heavy pelage. Such animals, therefore, have alternate routes for venous return which avoid close proximity to the arteries of the extremities. They can, therefore, adjust to fairly wide extremes of environmental temperature by selecting the appropriate route for venous return of blood from the extremities.

Both birds and mammals avoid low temperatures by undertaking long migrations. Other tolerate decreased body temperature by entering a state of torpor or hibernation. Newly hatched or newborn vertebrates are generally less capable of thermoregulation than are older animals of the same species. They are smaller, have less pelage, and have immature physiological responses. Appropriate nests and parental care are provided as necessary, and breeding places and times are selected so that the young will have the benefit of the best season and climate for their early growth and development.

Adult birds and mammals avoid temperature extremes in nests and burrows, and they curl up to reduce exposed skin area or huddle with others of their own kind to preserve body heat.Cooling generally requires water evaporation and, therefore, adaptations to control body water loss as well. Evaporative cooling is involved in sweating, of course, and in the panting activity of many mammals. Birds engage in gular fluttering, an activity much like panting. Some mammals that do not sweat (e.g., rodents) wet the fur by licking it and thus achieve the same end. In addition, bird and mammals avoid high temperatures by burrowing, or by seeking shade or damp ground to lie on. Birds sometimes fan themselves with their wings or, under certain conditions, dust themselves to improve the thermal conductivity through the feathers. Increased vascularity of the skin promotes heat loss by radiation and convection as well.

In warmblooded animals these receptors, operating through nerve reflex arcs, induce shivering (heat production), sweating (heat loss), and other thermoregulatory responses. In cold-blooded animals similar thermoregulatory responses occur but they take the form of behavioral adjustments. The animal moves into the sunlight or the shade to avoid dangerous environmental temperature extremes. The location of internal temperature receptors is not known, but it is probable that they are within the central nervous system itself and respond to the temperature of the blood circulating there.

Humans and many other higher vertebrates have both warm and cold receptors present in the skin, especially in that of the tongue. These appear to be finely branched nerve endings which probably respond to changes in their own metabolic rates rather than to air temperature or infrared radiation directly. Gradations from cold through cool, warm, and hot involve differential responses of cold receptors, warm receptors, and pain receptors.

Chemoreception

Lower vertebrates in aquatic environments retain this direct sensitivity in the surface cells of their skins, and this gives them a general but rather indiscriminate means of sampling composition of the surrounding water. Simple chemoreception haves been insufficiently studied but it appears to operate primarily to detect and avoid harmful substances. The receptors are not very sensitive; they detect only rather sharp changes in the immediately adjacent environment. Some fishes have, in addition to the general chemoreceptivity of skin cells, specific taste receptors scattered in the skin. These are most numerous in the head region of the animal. They have greater sensitivity than the receptors just described and can probably discriminate between different chemical substances to some extent. Still more sensitive and more discriminating are the olfactory receptors which occur also among fishes. The development of more responsive chemore-ceptors makes it possible for these animals to detect chemical substances in very low concentrations in the water, and, therefore, at some distance from their sources.

Perhaps paralleling the development of olfactory senses was the evolution in various species of fishes of skin glands that produce species- and sex-specific chemical substances. The odor of these materials permits recognition of friend, foe, or potential mate at some distance. In non-aquatic vertebrates the taste cells remain as contact-sensitive cells with poor discrimination and are generally further restricted in locality. Wet-skinned amphibians retain some taste cells in the skin of the head, but higher vertebrates limit the distribution of these receptors to the tongue and palate regions. Non-aquatic vertebrates also retain olfactory receptors of great sensitivity and discrimination and can, therefore, detect air-borne odors at some distance from the source.

Taste reception

Human taste bud contains about twenty modified epithelial cells grouped together. It is estimated that adults have about 10,000 such taste buds, mostly located on the tongue. New cells constantly arise at the periphery of each taste bud and migrate toward the center where, as old cells, they die and are removed. From each taste cell project several microvilli which extend through a central taste pore in the overlying epithelium. The branches of a nerve ending lie between the cells of the taste bud. Each taste bud contains cells which are particularly sensitive to one specific taste modality, although each can be stimulated by any modality if the stimulus is sufficiently intense. There is no detectable structural difference in the appearance of taste buds having different sensitivities so the basis of modality discrimination is not known.

A physiological difference in the buds themselves apparently exists, however, since the application of cocaine abolishes sensitivity to taste modalities in a definite sequence. Sensitivity to bitter disappears first, then sweet, salt, and finally sour these are the four modalities that human taste buds can differentiate. Some authorities believe that a fifth modality, metallic, is also separately discriminated by taste buds; others contend that the metallic flavor is a combination of the basic four modalities plus, perhaps, an ionic effect stimulating touch receptors.

Taste buds that are particularly sensitive to each of the basic four taste modalities are localized in different areas of the human tongue. Sweet-sensitive taste buds are located mostly near the tip of the tongue, while bitter-sensitive taste buds occur almost exclusively near its base. It is a matter of common experience that the sensation of bitter taste is a kind of aftertaste; the bitter substance has already been well chewed or, perhaps, swallowed- before one becomes conscious of the bitterness. This is because the substance is not tasted until it has contacted the root of the tongue where these receptors are located.

Sour-sensitive taste buds occur primarily along the lateral margins of the tongue; sensitive taste buds are more uniformly scattered over the entire surface. Sweet taste sensations are brought about by sugars, a number of different alcohols, aldehydes, ketones, esters, amino acids, and several other organic compounds as well as some inorganic salts. Many substances evoke stronger responses from the sweet receptors than does common table sugar. For example, nitrobenzene is about 5000 times sweeter than sucrose. The taste threshold for sucrose is about 0.01 molar; that is, a solution of sucrose need be only that concentrated to be detectable as a sweet taste to the average human.

Bitter tastes are also characteristic of a number of substances, most of which are long-chain organic compounds or alkaloids. Alkaloids are complex organic materials found in plants, many of which have ph, logical effects on animals and are used in medicine. Some substances that taste sweet at first also have a bitter aftertaste because they stimulate the sweet receptors at the tip of the tongue first and only later the bitter receptors at the base of the tongue.

Olfaction

Olfaction operates as a distance receptor only because it is some 10,000 times more sensitive than taste. Thus, the substance to be detected can be present in much lower concentrations, i.e., in the concentrations present in the air (or water) at a distance from the substance. The scientific study of olfaction is fraught with difficulties and much less is known about it than, for example, about vision or hearing. There are two major reasons for this. Olfaction has its central nervous connections in the phylogenetically older, more primitive portions of the brain. The effects of olfactory stimulation often involve subconscious emotional attitudes and basic social 'behavior, which do not often lead to predictable modifications of activity on a conscious level, i.e., activity that can be observed quantitatively. Scents are used by animals in their search for food, for species and sex recognition, for marking the boundaries of territories, and for other fundamental social purposes. Use of scent in this way reaches its peak of development among social insects. Individuals of an ant or bee colony band together as a sort of superorganism within which olfactory substances called phaeromones act to coordinate the activity of the colony as a whole. The situation is almost perfectly analogous with the role of hormones in coordinating the activity of the many cells banded together to form a vertebrate animal.

For humans the odors of good perfume, new automobiles, libraries, or dental offices are tied to emotional attitudes derived through past experience. These odors evoke subtle behavioral patterns rather than producing immediate, overt response patterns that can be easily studied. Humans are capable of identifying separately hundreds of different substances by their odors. It seems certain that there cannot be separate receptors for each of these. Rather, most of them must be combinations of a relatively few primary odors just as certain colors result from the mixture of primary colors.

Attempts have been made to identify the primary odors and some success has been achieved. It now seems likely that humans differentiate seven primary odors: camphoraceous, musky, floral, minty, ethereal, pungent, and putrid. Studies of frogs have yielded evidence that they distinguish five of the same modalities plus three additional primary odors not discriminated by humans. Except in the cases of pungent and putrid odors, the molecular conformation of an odor-emitting substance probably enables it to "fit" a specific receptor surface and, thereby, stimulate that receptor. Thus, any substance; with similar molecular shapes will be perceived as having the same primary odor.

Pungent and putrid odors, on the other hand, are apparently distinguished on the basis of the molecular charge. Combinations of more than a single odor are probably perceived as the algebraic sums of the primary odors present in the mixture. Commercial deodorizers presumably work only because they combine with and, therefore, mask the molecular shapes of the odorous materials present in the air. It has been recognized for some time that certain kinds of chemical structures are associated with strong odors and that relatively minor chemical adjustments of the molecule could change the character of the odor without interfering with its strength.

The human olfactory apparatus lies in the upper, back, and lateral portions of the nasal cavities. There are about 10 to 20 million nerve endings there which cover an area of about five square centimeters. The receptor cells are actually nerve cells which are modified by the possession of sensory hairs or cilia. The hair surfaces are, presumably, the receptor sites for odorous substances; if this is so, there is an enormous receptor surface available. Although human olfaction is apparently somewhat less sensitive than that of most mammals, the thresholds are surprisingly low.

The olfactory pits of teleost fishes are located dorsally on the snout and generally are not connected with the respiratory system. In some cases, however, the same muscular activity forces water through both the gills and the olfactory pits. Typically the olfactory pits have separate incurrent and excurrent openings, and water flows through them because of respiratory movements, ciliary activity within the pits, or forward swimming through the water.

In the lungfishes the olfactory cavaties open dorsally on the surface of the snout and ventrally into the oral-pharyngeal cavity, as they do in all higher vertebrates. Fishes apparently make great use of olfaction and have very good sensitivity and discrimination. It has been shown conclusively, for ample, that salmon return to the precise stream in which they hatched, presumably because of olfactory memory. The differences in dissolved materials in two adjacent tributaries of a stream must be exceedingly small, yet there is sufficient difference to allow the salmon to make the proper choice. Sharks are notoriously sensitive to the presence of even small amounts of blood in the water. They detect it at great distances and quickly gather at the site.

The olfactory organ of reptiles are called Jacobsen's organ. This organ consists of a pair of pits opening into the roof of the mouth. The typical forked tongue is protruded to sample the air and then withdrawn and inserted into the openings of Jacobsen's organ. Thus, chemical materials from the air are placed in direct contact with the olfactory sensory cells. Since reptiles generally require less respiratory ventilation than other vertebrates, this adaptation probably is an improvement over the dependence upon respiratory passage of air over the olfactory epithelium.

Sense of smell is very poorly developed in birds; olfaction is of little use in the sky. Vision is of much greater importance and the great development of visual acuity in birds has led to increasing dependence upon that sense to the virtual exclusion of olfactory development. Birds must keep the "payload" at a minimum and cannot, therefore, afford to carry organs that arc of little use, especially when other organs must become more highly developed and more efficient, Hummingbirds select flowers on the basis of color alone and are attracted to a piece of red cloth as strongly as to a red flower. Most species of vultures will ignore a well-ripened carcass if it is covered and, therefore, not visible to them. Some vultures, which do presumably have a sense of smell, are attracted to leaks in cross-country natural gas lines. Gas line inspectors watch for accumulations of vultures along the lines as a means of locating leaks. Apparently, to the vulture, methyl mercaptan smells like rotting flesh. The kiwi is another exception to the rule. This flightless bird locates worms beneath the surface of the soil by olfactory means.

Mechanoreception

Cutaneous mechanoreception

Mechanoreceptors are merely free nerve endings. Still, in such cases the stimulus induces a receptor potential in the free nerve ending much like the generator potential of more complex receptors. This is not conducted but evokes true nerve impulses arising at the first node of Ranvier when the receptor potential exceeds the firing threshold at that node. Mechanoreception in its simplest form is a very primitive sense, one that is perhaps as old as life itself. Human skin has a total of nearly three quarters of a million discrete receptors. They include specialized structures such as pacinian corpuscles, Meissner's corpuscles, and, in a few places, end bulbs of Krause. They also include free nerve endings, some of which form rings about the bases of hair follicles so that mechanical disturbances of the hair cause stimulation of the associated nerve.

Cutaneous receptors are unevenly distributed, being, most concentrated in areas surrounding body orifices, the eyes, and the fingertips. The relative distribution can be tested by determining the two-point resolution in various skin areas of an individual. The test determines how far apart two stimuli must be in order to be distinguished as two, rather than as a single stimulus. The blindfolded subject is touched sometimes with one and sometimes with a pair of sharp points and asked to indicate whether he feels one or two separate stimuli. The distance between these points is slowly widened until the always gives the correct answer. Theoretically, in order to resolve them as two separate stimuli, each must produce a response in a different, single receptor.

Proprioception

Receptors located in skeletal muscles and in the associated tendons and joints are mechanoreceptors. They are stimulated, in most cases, by stretching of the tissues in which they lie. Three discrete receptor types are recognized and together they continuously inform the central nervous system of the current position and Movements of body parts. Joint receptors are located within the joint capsules of the skeleton. When the joint is moved the receptors are subjected to stretching, which is the effective stimulus to them. Probably each is stretched by movement of the joint in a given direction only but collectively they monitor all possible joint movements. They respond in a tonic manner, indicating both the rate of movement and the final position of the joint. Golgi tendon receptors are also tonic receptors sensitive to stretching of the tendon.

An additional burst of action potentials in the associated nerve encodes the rate of change, and a subsequent steady state of discharge frequency indicates the final static degree of stretch. Two different kinds of activity may result in the stretching of a tendon, and the Golgi receptors cannot distinguish between the two: (1) contraction of the associated muscle stretches its tendon and (2) contraction of the antagonistic muscle also stretches it. For example, contraction of either the biceps brachii muscle or of the triceps brachii muscle will result in stretching of the tendons of both muscles.

Together with the input from the joint receptors of the elbow, however, the central nervous system is given sufficient information to determine which of the two muscles contracted and how rapidly that contraction occurred. It will be recalled too that after the biceps contracts, a reflex inhibition of the triceps soon occurs. This, undoubtedly, reduces the stretching of the triceps tendon while that of the biceps continues. Muscle spindle fibers are actually modified muscle fibers lying here and there within skeletal muscles. Each spindle fiber has an independent motor nerve supply so that it can contract or relax independently of the surrounding muscle fibers. Only the ends of the spindle fiber are contractile and these are innervated by this motor nerve.

The central portion of the spindle fiber is a sensory receptor and is innervated instead by a sensory nerve. The central portion responds in a tonic fashion to the stretch imposed upon it by the surrounding muscle fibers. It is stretched, specifically, when the surrounding fibers relax, since relaxation of a muscle increases its length. When the surrounding muscles contract, the degree of stretch of the spindle fiber is lessened. Similar to a typical tonic receptor, the spindle fiber responds to changes in the degree of stretch by inducing a burst of nerve impulses at a frequency that is proportional to the rate of change of stretch. It then settles down to a frequency level that is proportional to the final degree of stretch.

Unlike many tonic receptors, however, the spindle fiber follows this reaction with an adjustment to a base frequency which is always the same. For example, if the surrounding muscle relaxes, the spindle fiber is stretched. An initial overshoot of generator potential encodes the rate of stretching in an initial burst of nerve impulses. Following the overshoot, the generator potential levels off, and nerve impulses proportional to the static degree of stretch reach the central nervous system. The contractile ends of the muscle spindle relax enough to exactly relieve the stretch, and the tonic frequency of impulses in the associated sensory nerve drops to a standard resting level. The reverse occurs when the surrounding muscle fibers contract. Thus, these receptors inform the central nervous system of the rate and amount of change in muscle contraction and then "reset" themselves so that they are ready to respond in a like manner to the next change in muscle contraction. Whereas muscle spindles monitor only briefly the final degree of muscle contraction, the tendon receptors continue to monitor it, and, it will be recalled, the joint receptors indicate the direction and rate of resulting joint movement as well as the final position of the joint.

Internal mechanorecptors

Mechanical interoception is found in the hydrostatic pressure receptors of the carotid sinus and the arch of the aorta. These are tonic receptors that constantly monitor the level of arterial blood pressure. The walls of swim bladders of fishes contain similar receptors. Another example of internal mechanoreception is found in the stretch receptors of mammalian lungs. They are also tonic receptors which monitor the degree of inflation d the lungs in a continuous fashion. Related receptors occur in the pharyngeal walls of some fishes where they play an analogous role in the control of gill ventilation. There are other internal mechanoreceptors, including pacinian corpuscles, that are apparently identical with those of the skin. A very large corpuscle occurs in the mesentery of the cat and it has been the object of much investigation because of its size and location. Usually there is no conscious awareness of the central nervous input from internal mechanoreceptors; this input functions instead through reflex mechanisms at subconscious levels.

Lateral line organs

Fishes, amphibian larvae, and some amphibian have rows of sensory receptors called neuromasts along their sides. These, collectively, comprise the lateral line organ. The lateral line organ is generally marked by associated pigmentation so that its location is visible as a single stripe running length wise along either side of the animal. Each neuromast is a cell having a hairlike extension that is embedded in a gelatinous mound or column called a cupola. Sometimes the receptor cells lie in the bottom of a shallow groove that opens to the side of the side of the animal, but more frequently the groove is covered over, thus forming a canal lying near the surface. These canals open to the surface at intervals. Lateral line canals occur particularly in more active forms, whereas the open groove system occurs in certain more sedentary forms. The primitive function of neuromasts seems to have been as rheostatic receptors, i.e., receptors sensitive to currents in the water. Water current, by displacing the cupola of the receptor cell, stimulates the embedded receptor hair.

Lateral line systems seem to contain two types of receptors, and water currents moving from the head toward the tail of the animal stimulate some receptors but inhibit others. Thus, it is possible for the fish to detect the direction as well as the comparative velocity of the water current. This kind of sensory system enables a fish to swim upstream at a velocity exactly matching the velocity of the downstream movement of water so that the fish is able to maintain its position with respect to the bottom, or to actually move upstream or drift downstream at appropriate rates. Thus a fish is able to detect its "water speed" by means of the lateral line, system and its "ground speed" visually a situation somewhat analogous to the simultaneous estimation of air speed and around speed by birds.

The same lateral line receptors are also able to detect disturbances in the water which produce deviant currents. A wriggling worm might produce such currents, and, moreover they would be repeating currents (or pressure waves) of low frequency. If the lateral lime system were to attain greater sensitivity, it could presumably detect such disturbances at greater distances and there is no reason why it might not attain the ability to detect higher frequencies (shorter wave lengths) resulting from such disturbances. Such sensitive reception would constitute "touch at a distance" or, ultimately, the detection of under water. Sound of relatively low frequency.

Thus, the lateral line system is fundamentally capable of developing as an organ of hearing. This has not actually occurred in the lateral line system itself, although some fishes can detect vibrations up to a frequency of about 100 vibrations per second through lateral line reception. Derivatives of the primitive lateral line system have developed in vertebrates as organs of hearing, and with these humans can detect frequencies as low as about 10 vibrations per second. The distinction between the detection of vibration by the lateral line and the true reception of sound is, therefore, somewhat artificial. Because movement through motionless water creates an apparent current and because this varies on both sides of the animal as it alters course, it is clear that the lateral line organ is fundamentally capable of monitoring linear (straight-line) and angular (turning) acceleration and deceleration. The structures of the vertebrate inner ear contain apparatus for monitoring linear and angular acceleration and deceleration, and these structures were also undoubtedly derived during evolution from portions of the lateral line system.

Photoreception

The visual receptors of vertebrates are variously adapted to meet the requirements of different species in a variety of ways. Frequently such special adaptations mean the striking of a compromise between two opposing situations. For example, sensitivity to low levels of light requires the additive effects of several receptors, a form of spatial summation. This, in turn, means a reduction in two-point discrimination and, therefore, only rather crude image formation. The formation of a fine-grained, detailed image requires good lighting and the differential response of many individual receptors. Thus, nocturnal and diurnal animals have different and opposite requirements in this regard.

Those animals which are active both day and night (e.g., humans) must strike some compromise that is adequate in both situations. It is pointless for a fish to have great distance vision since the light-transmitting qualities and the turbidity of most natural waters are limiting factors. For a high-flying predacious bird such as an eagle, however, such distance vision is essential. Moreover, the optical qualities of water and of air are quite different, so the structure of the eyes of fishes and eagles must be quite different in still other ways.

A very wide visual field is ordinarily possible only with non-stereoptic or monocular vision, a type which is advantageous to an animal that is preyed upon. Such an animal merely needs to detect the presence of the predator, in whatever direction it may be and move away from it. Thus, the eyes of a cat are placed anteriorly where the visual fields can broadly overlap to give stereoptic vision straight ahead, whereas those of a rabbit are placed laterally where they can together survey the entire horizon.

Animals, such as predacious birds, that need both generally searching and localized attacking vision must compromise in some way between these extremes. Similarly, the ability to perceive motion, color, and contrast is variously present or absent in different vertebrates. The presence of such adaptations corresponds more closely with the ecology of the animal in question than with its phylogeny, a fact which demonstrates that similar adaptations have arisen independently many times in vertebrate evolution.

The human eye

Human eye is a nearly spherical, fluid-filled, hollow organ. Its walls are composed of three general layers variously modified in different regions of the eyeball to perform different functions. The outermost layer is composed of tough conn-ective tissue that gives rigidity to the spherical body of the eye. This portion, called the sclera, is modified ante-riorly to form the optically clear cornea. The corneal bulge has a greater curvature than that of the sclera and is closely covered by a thin conjunctiva. The latter is a transparent continuation of the skin of the face which also lines the spaces behind the eyelids. Together, the cornea and the conjunctiva form the only air-tissue in face in the vertebrate optical system and as the chief refracting surface of the eye. The innermost layer of the eyeball is the retina, it is present only ton the sides and the posterior portion of the eyeball.

The retina is, of course, the layer containing the photoreceptor cells about which we shall have more to say presently. Between the sclera and the retina is an intermediate vascular layer called the choroid. Besides carrying the major blood supply to structures of the eye, this layer contains on its inner surface a heavy deposition of pigment. Continuous with the choroid anteriorly are two sets of muscles. One of these forms the iris diaphragm, the colored portion visible in the eyes of others. It is composed of concentrically and radially arranged muscles which control the size of the pupil of the eye. This, in turn, controls the amount of light entering the eye. Just behind the pupil is the crystalline lens of the eye. This is held in position by suspensory ligaments which are anchored into the connective tissue framework (sclera) of the eyeball. Also attached to the suspensory, ligaments are fibers of the ciliary muscle. This, the second muscle group, is more or less continuous with the choroid layer of the eyeball. These muscles function in the control of the curvature of the crystalline lens. The anterior and posterior chambers of the eye lie on either side (anterior and posterior) of the iris diaphragm and are both filled with a fluid called the aqueous humor. The larger cavity, which is posterior to the crystalline lens, is filled with a semisolid transparent mass called the vitreous humor. The retina is thinnest at its edges and becomes progressively thicker toward the posterior and medial Point, which is known as the optic disc. At the optic disc the retina becomes continuous with the optic nerve. The optic nerve is covered by a connective tissue sheath, which is a continuation of the sclera, and passes posteriorly along the base of the brain.

At the posterior pole of the retina there is a slight depression, or thin spot, known as the fovea centralis. This is the area of most acute vision. It is surrounded by a raised, thickened region called the macula lutea. Attached to the sclera externally are six skeletal muscles which rotate the eyeball in various directions. Another extrinsic muscle, the levator palpebrae, lifts the eyelid to open the eye. The iris diaphragm is the pigmented ring, which may be black, brown, blue, or, less commonly, another color, surrounding the pupil of the eye and situated just anterior to the crystalline lens. It is composed, as we have seen, of two sets of muscle fibers. One set is arranged concentrically and by contracting decreases the diameter of the pupil. The other set of muscles radiates in all directions from the edge of the pupil to the edge of the iris. Contraction of these muscles increases the diameter of the pupil. Thus, the relative degree of contraction of the concentric muscles and the radiating muscles determines pupillary aperture at any given time. These two sets of muscles are controlled by the autonomic nervous system.

Sympathetic stimulation enlarges the pupil while parasympathetic stimulation constricts it. Pupillary size plays two important roles in vision. Most obvious is the limitation of the total amount of light entering the eye. Light glare on the retina acts through reflex nervous pathways to cause pupillary constriction. Dim light acts through similar pathways to cause pupillary expansion. In addition, the pupillary aperture is important in determining the depth of focus of the eve and must be adjusted when the eye views objects that are close to it. The human pupil is adjustable between diameters of 1.5 and 8.0 millimeters. The amount of light entering the eye is proportional to he square of the diameter, so this range permits a 30-fold difference in the light intensity reaching the retina.

When objects are sufficiently far away from the eye, light rays reflecting from them enter the eye on essentially parallel courses. For all practical purposes this is true for any object which is slightly beyond arm's reach. The refractive power of the air-tissue interface plus that of the crystalline lens in its natural resting shape is such that the images of distant objects fall exactly upon the retina. When an object is moved closer to the eve, however, light rays reflecting from it enter the eye on divergent courses. If no adjustments were made to correct for this, the image would fall further behind the lens. that is, actually behind the retina.

If the refractive power of the eye could be increased in some way, the plane in which the image is formed could be moved forward so that it fell exactly upon the retina as it should. This can be accomplished by increasing the curvature of the crystalline lens. Relaxation of the suspensory ligaments is accomplished by contraction of the ciliary muscles of the eye. The fibers of these muscles are arranged in such a way that contraction pulls the scleral attachments of the suspensory ligaments forward and centrally toward the crystalline lens. This lessens the pull of the suspensory ligaments on the crystalline lens and allows the lens to become thicker and more refractive.

Thus, while the air-tissue interface at the surface of the cornea and the conjunctiva is the greatest refractive surface of the eye, changes in shape of the crystalline lens are sufficient to make the small adjustments in refraction that are necessary to accommodate the eye to near vision. Because close work requires constant contraction of ciliary muscles, such work is tiring to the eyes. Distant vision requires no such muscular activity and can be carried on for prolonged periods without noticeable fatigue. With increasing age the crystalline lens tends to lose much of its elasticity.

As a result, contraction of the ciliary muscles has little or no effect on the shape of the lens. It becomes increasingly difficult for persons older than about 45 years to focus accurately upon objects held close to the eyes, and they commonly tend to bold reading matter as far away as they can while examining it. This condition, known as presbyopia, frequently requires the use of reading glasses or, if glasses are already worn for the correction of other defects, the addition of a bifocal lens to the eyeglasses. Bifocal lenses have an upper region for distance viewing and a lower region, which is more refractive, for reading and other close vision. When an object is held close to the eye, a greater proportion of the total amount of light reflected from that object is able to enter the eye.

Retina of the human eye contains receptor cells of two types, rods and cones, as well as associated neural cells and their nerve processes. Each human retina contains about 1.2×10^8 rods and about 5.5×10^6 cones. The nerve cells and processes are innermost so light must penetrate through these layers before reaching the sensory cells themselves. This seemingly inefficient arrangement is a consequence of the embryonic formation of the retina as an extension of the central nervous system. The area within the fovea centralis of the retina contains closely packed cones but no rods. The retina is thinner at this point and thicker in the immediately surrounding area because the neural elements are diverted to one side. In this region, then, light penetration to the receptor cells is greatly improved.

From the fovea centralis outward in all directions there are progressively fewer cones and more rods until at the periphery of the retina only rods occur. Both rods and cones contain photosensitive chemical substances which undergo reactions when they absorb photic energy. Much less is known of the chemistry of cone reception, but it is undoubtedly similar in principle to the chemistry of the rods, about which great deal is known. In the region overlying the optic nerve there are no receptor cells at all. Consequently, this part of the retina called the optic disc, is blind.

Cones require good light intensity and are, therefore, only stimulated in daylight (photopic vision). Each cone is associated with a separate neuron leading to the visual cortex on the occipital lobe of the cerebrum; this guarantees good resolution, or two-point discrimination. Cones are capable of distinguishing color, whereas rods are not. Again, if one looks fixedly at an object and then shifts his attention to the side, he will find that colors are seen only in or near the center of the visual, field. This is true because the peripheral portions of the visual field fall upon peripheral portions of the retina where only rods are present. The rod cells are relatively sensitive and require only small amounts of light to be stimulated. This sensitivity permits scotopic or night vision, and it is a matter of common experience that in darkness one sees the periphery of the visual field better than the central portions. The sensitivity of the rods is probably related to the fact that several rods typically converge upon a single neuron. Such an arrangement makes spatial summation possible, so that low intensities of stimulus are sufficient to produce action potentials in the common neuron. At the same time, however, resolution cannot be as good because several receptors effectively serve as a single receptor.

Night vision, therefore, lacks the precise two point discrimination of day vision. Also, of course, night vision lacks color discrimination since the rods are incapable of it. The structure of rods and cones is highly specialized. Rods have a terminal portion that is apparently derived from a cilium. The nine typical peripheral fibers of a cilium are present but, in addition, there are many disclike plates having what appears to be a unit membrane structure. These plates also contain a specific photoreactive substance called rhodopsin. The adjacent segment of the rod cell contains numerous mitochondria. The terminal segment of each rod contains the photosensitive pigment rhodopsin, or visual purple as it is sometimes called. When this substance absorbs a sufficient amount of light energy, it splits into its constituent parts, opsin and trans-retinene. This reaction alters the generator potential of the rod, which, in turn, produces activity of the associated neuron. Rhodopsin can be resynthesized by recombining the retinene with the opsin but this recombination requires the prior isomerization of trans-retinene to cis-retinene. The enzyme retinene isomerase catalyzes the reaction. Both trans- and cis-retinene can be derived from the corresponding isomers of vitamin A. This relationship explains the important role of vitamin A in night vision.

Similar photochemical processes take place in cones, and one photosensitive substance, iodopsin, has been identified there. In order for color discrimination to take place it is postulated that three different types of cones must be present. Each type, theoretically, contains a different photosensitive compound which absorbs light within a different range of wave lengths. One of these absorbs red light, one absorbs green light, and one absorbs in the blue-violet portion of the spectrum.

Nocturnal animals tend to have larger corneal surfaces than do diurnal forms. This results in more efficient light-gathering properties and, therefore, more acute night vision. The eyes of cyclostomes and teleosts are generally focused for near vision at rest and must be actively accommodated for distance vision. It will be recalled that this is the opposite of the situation in humans. Accommodation for distance vision in these fishes is accomplished not by a change in the lens shape as in humans, but by a change in its distance from the retina. A special muscle pulls the lens backward to shorten the lens-to-retina distance. Conversely, the eyes of elasmobranchs, amphibians, and snakes are focused at rest for distance vision and must be actively accommodated for near vision.

Members of all vertebrate classes, have increased the total illumination of the retina by the presence of light reflecting substances in the pigment layer of the choroid. Usually the tapetum lucidum, as the reflecting layer is called, is composed of crystalline guanine, but in some instances other different reflective materials occur. In any case, while reflection increases the total illumination and, therefore, the sensitivity of the retina, it does so at the cost of good resolution. The tapeturn lucidum is responsible for the "nightshine" seen at night in the eyes of many animals, including cats and dogs, when their eves are illuminated by the headlights of all automobile or by a flashlight.

In elasmobranchs black pigment cells mask the tapetum lucidum in daylight but concentrate their pigment to expose it at night. Thus, the eye is capable of good resolution in daylight and heightened sensitivity in the dark. Such pigment migration also occurs in a number of other submammalian vertebrates. In other cases, the pigment is unmoving but the receptor cells themselves move into or out front the pigment layer thus accomplishing the same ends. In general, such changes occur rapidly in lower animals, more slowly in higher forms, and not at all in mammals.

During vertebrate evolution there has been a progressive development of pupillary responses as an alternative and better mechanism for controlling retinal illumination. Only a few fishes have variable pupil diameters and in these the response is slow. In these fishes, apparently, the iris responds directly to light intensity, whereas in higher forms the response, which is fast and effective, is mediated by way of nervous reflex arcs. The slit-shaped pupils characteristic of many reptiles and some mammals can be closed more completely than those which are round, as is the human pupil. This is an advantage in an eye that is highly adapted to night vision but capable of some day vision with subdued retinal illumination.

Some animals, particularly birds, utilize colored filters in various ways. Tinted lenses or colored on droplets overlaying the retinal cells can act as ultraviolet filters or haze-piercing "sun glasses." These are of great advantage especially to high-flying birds, In some cases retinal on droplets of red, orange, yellow, or even green occur which apparently function in food gathering. Some birds (e.g., hawks, hummingbirds) must be able to search in all directions by the independent use of both eyes (monocular vision) and also to be able to "attack" using the depth perception that comes from the simultaneous and coordinated use of the eyes (binocular vision). In such cases, each eye contains two separate foveal areas, one central fovea for monocular vision and one temporal fovea for binocular vision.

Birds which float on the water and other vertebrates which live on open plains often have foveas that are extended, horizontal strips, corresponding with the image of the horizon. In some birds the retina has greater cone density in one half of its area than in the other half. For example, Acipiter acipiter, the goshawk, is a high flying bird which normally hunts ground-dwelling prey. The top half of the retina receives the image from the ground below in great detail because of the concentration of cones in that half. The lower half of the retina has such poor resolution that, when the bird is perched, it must turn its head upside down in order to inspect the sky above.

In the eyes of both terrestrial and marine vertebrates, a characteristic photosensitive pigment called rhodopsin is present. Rhodopsin is composed of opsin plus the 11-cis isomer of retinene, a derivative of vitamin A. In fresh-water fishes, however, the visual pigment of the rods is called porphyropsin and contains the 11-cis isomer of retinene a derivative of vitamin A.

Among the amphibians which are primarily terrestrial as adults, the pigment changes from porphyropsin to rhodopsin at the time of metamorphosis, when the animal leaves the freshwater habitat. Fishes that migrate between fresh and marine waters have mixtures of the two pigments but the ratio shifts in the appropriate direction during the migration-more rhodopsin when in the sea and more porphyropsin when in fresh waters. The two pigments have different maximum absorption wave lengths, rhodopsin being better adapted to receive the shorter wave lengths of light. The opsin portion of either rhodopsin or porphyropsin also has a great influence on the wave length of maximum light absorption. In marine fishes, for example, greater habitat depths are generally associated with rhodopsins, which absorb shorter wave lengths. These short wave lengths, in the blue-violet end of the spectrum, penetrate deepest into the sea. Many organisms which live on the sea bottom are bright red or orange in color, and one would think that this would make them apparent to predators.

Audio reception

During the embryonic development of vertebrates, an ectodermal sac called the otic vesicle forms on each side of the head. The lining of the otic vesicle contains certain tissues which in the lower aquatic vertebrates include parts of the lateral line organ. In adult vertebrates the otic vesicle becomes the inner ear and contains sensory receptors that are derived from neuromasts like those of the lateral line organ. If receptors of this general kind, then, could be modified so that their sensitivities were limited to a single modality from this list, they would be far more useful to the organism. Unrestricted response to each of a variety of stimulus modalities yields less precise information to the central nervous system. Limiting some neuromasts to one modality and others to a different modality has occurred in the vertebrate inner ear.

The modifications are brought about by isolating them where extraneous stimuli are unlikely to reach them and by surrounding them with specialized accessory structures that serve to emphasize the effect of single modalities. In mammals the inner ear receptors lie within a cavity in the temporal bone of the skull, a cavity called the bony labyrinth. This cavity contains a series of organs composed of soft tissues collectively spoken of as the membranous labyrinth. The membranous labyrinth does not completely fill the bony labyrinth and the remaining space is filled with a fluid called perilymph. The membranous labyrinth itself is filled with a similar fluid called endolymph. The endolymph is somewhat more viscous than the perilymph, and there is no point at which the two fluids are in direct contact with each other, they are separated everywhere by membranes of the membranous labyrinth. The coiled cochlear portion is the organ of hearing in mammals, which does-not occur in lower vertebrates.

The remainder of the mammalian inner ear is called the vestibular portion and contains receptors responding to linear and angular acceleration-deceleration, gravity, and perhaps low-frequency vibration.

Semicircular canals

Semicircular canals of vertebrate ear is half circle shaped organs. It is composed of a tubular extension of the membranous labyrinth lying, within a larger tubular extension of the bony labyrinth. Each membranous tube is attached at both ends to a central membranous sac called the utricle. All higher vertebrates have three semicircular canals on each side. They are oriented in planes that are all at right angles to each other, so there are two vertical canals and one horizontal one. The horizontal canal is the one missing in the cyclostome ear.

At one of the ends of each membranous semicircular canal is an enlargement called the ampulla. This ampulla is the site of the receptor organ, a structure called the crista. The crista closely resembles lateral line neuromasts, being composed of cells that have hairlike extensions embedded in a gelatinous cupola. The cupola extends completely across the cavity of the ampulla so that any movement of endolymph in either direction moves the cupola backward or forward like a swinging door. Such movement results in either raising or lowering the generator potentials of the hair cells, depending upon the direction of movement. Since they are tonic receptors, this results in an increase or decrease in the frequency of nerve action potentials to the central nervous system.

The function of the semicircular canals can perhaps, be best understood by means of an analogous model. If a bucket of water is suddenly rotated about its own axis through a part of a rotation, the water within the bucket tends to remain at rest because of its own inertia. The bucket, in effect, rotates about the water while the water remains still. If a structure like the cupola were attached to the inside surface of the bucket so that it extended into the water, such a movement of the bucket would cause deflection of the cupola in the opposite direction. Actually, the cupola would be moving through the water rather than the water flowing past the cupola. Nevertheless, the cupola would respond to this apparent current just as it would to a real current in the water. Rotation of the human head about its own axis produces apparent currents of the same kind in the horizontal semicircular canal of each inner ear. There is not much actual fluid motion because endolymph is a fairly viscous fluid, but it is sufficient to cause deflection of the cupola and, consequently, a change in the frequency of action potentials in the associated neurons

Since semicircular canals lie in the other two dimensional planes as well, rotation of the head about any other axis will also produce apparent currents in at least one of the three semicircular canals. Rotation in most directions will, in fact, produce such apparent currents in two of the canals on each side. Such an event presents the central nervous system with input from both of the canals involved and, therefore, with a problem in vector analysis. The comparative amounts of cupola deflection in the two canals encodes all of the information necessary for the central nervous system to determine the true direction and speed of head rotation. The code, of course, is the frequency nerve action potentials coming from the two receptors, since these frequencies are proportional to the generator potentials established in those receptors.

Such rotational movement of the head constitutes angular acceleration (starting the movement) and deceleration (stopping the movement). Of course, the central nervous system also receives data regarding such angular or rotational movement from other receptors as well. The skeletal muscles producing the movement are equipped with proprioceptors; the effects of head movements are also monitored by the eyes. Such reductant sensory inputs reduce nervous system. In this particular case, however, such misleading and conflicting inputs are quite common and familiar.

Following a period of prolonged rotation, one feels dizzy for a time as though one were still rotating. This is because the semicircular canal organs are still being stimulated. Return briefly to the analogy of the bucket of water mentioned earlier in this section. If we rotate the bucket of water continuously through several rotations and then stop it, we will find that the inertia of the water at rest has been overcome and it has begun to rotate with the bucket. When the rotation of the bucket ceases, inertia again causes the water to continue in motion. The sensory organ that we visualized as being attached to the inner surface of the bucket is then stimulated by an actual water current for a time until the water again comes to rest. This requires some time for water but much less time for the more viscous endolymph.

Thus, following prolonged rotation the organs of the semicircular canal continue to send information to the central nervous system indicating that rotation is continuing when, in fact, it has stopped. This causes the sensation of vertigo or dizziness. To a certain extent, the central nervous system acts on this false information. The eyeballs, for example, reflexly move back and forth, rapidly in one direction and more slowly in the other, in a characteristic pattern known as nystagmus. This is precisely the way in which the eyes would move if actual rotation of the bead were continuing. While rotating, a person looks fixedly at some object, following it with his eyes until the rotation causes it to drift out of the visual field. Then he quickly moves his eyes in the opposite direction to fix on some new object which is just coming into view.

The membranous labyrinths in vertebrates are differentiated as semicircular canals plus two saclike portions, the utricle and the saccule. The semicircular canals communicate, as we have seen, with the utricle, which, in turn, opens through an endolymphatic duct to the saccule. In non-mammalian vertebrates the saccule is extended as a small flap of tissue called the lagena. In addition, in reptiles, birds, and mammals there is a tubular element extending from the saccule. This is called the cochlear duct, or scala media, and is associated with the organ of hearing.

The utricle, saccule, and lagena all contain sensory organs that resemble each other. Each consists of a patch of sensory hair cells much like lateral line neuromasts. The sensory hairs are embedded in a gelatinous mass, which also contains grains of calcium carbonate. These particles, called otoliths, have a density that is about 2.5 times as great as that of the surrounding tissues and fluids.

Therefore, they react differentially to the force of gravity. If the head is tipped to one side, the weight of the otoliths causes bending of the sensory hairs and therefore initiates a change in the generator potential of the receptor. This, in turn, causes a change in the frequency of action potentials in the associated neurons. The central nervous system is thus informed of the rate and degree of head tipping as well as of the final resting position of the head after such movement takes place.

Again, information from the proprioceptors of neck muscles as well as from vision provides confirmation to the central nervous system. It is probable that the receptors of the utricle, saccule, and lagena also respond to linear acceleration and deceleration since such movements in any given straight line would also cause bending of the hair cells. In this case the greater inertia of the otoliths would be responsible.

Auditory apparatus

The human ear has been so reduced in size and mobility, however, that it is virtually useless in this regard. Its function is more decorative than to gather or localize sounds. Sound waves are alternate condensations and rarefactions of the air and they move in all directions from any vibrating source. Since they are waves in the air, they have variable frequencies (cycles per second), and these are inversely related to their wave lengths. Sounds with short wave lengths have high frequencies and are beard as treble notes. Sounds with long wave lengths have low frequencies and are heard as bass notes. Sound waves also differ in amplitude (loudness) and in timbre. Different musical instruments sound different even when they are playing the same note because of differences in timbre.

All sounds are, of course, complex patterns of wave lengths, and the resulting pattern is considered to be musical only when the various frequencies bear some interesting and pleasant relationship to each other. When these relationships are lacking, the pattern becomes discordant and unpleasant or, if the wave lengths present are totally random, it becomes pure "noise." The black-paper cone in the speaker of a phonograph or radio is caused to vibrate by electromagnetic forces; this vibration produces waves of corresponding frequencies, which spread through the air (or water) in all directions until they strike other objects. There they are either absorbed (damped) or reflected (echoed). When sound waves enter the external auditory meatus of the human ear, they impinge upon the tympanic membrane (ear drum), which closes the innermost end of the auditory canal. This membrane is a thin, cone-shaped structure somewhat like the paper cone in a radio speaker.

Sound waves striking it cause it to vibrate in resonance with the frequencies of sound present. Thus, all wave lengths of sound striking the tympanic membrane at any given time produce patterns of vibration in it. The point of the tympanic membrane cone is directed inward (medially) and is in contact with the handle of the first ossicle (malleus) of the middle ear. Another point of the malleus is attached to the wall of the middle ear cavity just beside the edge of the tympanic membrane. This point serves as a fulcrum on which the hone rocks in response to the vibrations of the tympanic membrane.

The malleus is immovably attached to the second ossicle, the incus, so the two bones move together as though they were one. The point of the incus articulates with the third ossicle, the stapes, by a movable joint. The stapes sits, like a piston in a cylinder in the, opening of the oval window, which is an opening through the bone of the skull leading from the middle ear cavity into the perilymphatic space of the inner ear. Movements of the incus move the stapes in and out within the oval window again, like a piston in a cylinder. Thus, air-borne sound patterns are finally translated into patterns of inward and outward movements of the stapes. Since the footplate of the stapes is only about a twentieth as large as the tympanic membrane, there is a total concentration of the power of motion resulting from this linkage of ossicles. Thus, the strength of the thrust of the stapes against the perilymph of the middle ear is greatly increased.

The handle of the malleus is constantly pulled inward (medially) by contraction of the tensor tympani muscle. This action maintains the tympanic membrane under some tension. Similarly, the stapes is constantly pulled outward (laterally) by the stapedius muscle. The tonus of both muscles is under reflex nervous control and can he tightened to damp very loud sounds or loosened to increase the sensitivity to very soft sounds. Thus, the central nervous system is able to "tune" the sensitivity of the organ of hearing by controlling the movability of the tympanic membrane and ossicle linkage of the middle ear. The eustachian tube connects the cavity of the middle ear with the pharynx. Air can pass through the eustachian tube in either direction to equalize the pressure on the two-sides of the tympanic membrane. Everyone is familiar with the damping of sound that occurs during a fast ascent or descent in an elevator or unpressurized aircraft. This damping disappears when one swallows or yawns, activities which cause a momentary opening of the eustachian tube.

Reproductive organs

Function of reproductive system is to produce new individuals of the same species. While the offspring produced by the animals are of the same species, and therefore have the same body pattern, they are not ordinarily carbon copies of their parents through special cells to next generation. A new individual is, to begin with, a single cell containing this information, then ultimately a new individual of the species. In general, the information borne by a single cell results in the production may result in the production of the specific body form and action, but variations occur in every feature.

In almost all vertebrates sexual reproduction is taking place. The basic reproductory units are the gametes. Gametes are usually of two classes: sperms and eggs. Sperms generally are tiny and motile, usually moving about by means of a flagellum, but in some species having other means of propulsion. Eggs are not motile and are generally large as cells go. The gametes are not typical cells, for they have only half the number of chromosomes found in adult cells.

Mammalian reproductive organs

Male reproductive organs

The male reproductive organs in mammals are the testes. They are composed of cords of cells arranged in lobules. Spermatozoa arise from these cords of cells under the influence of FSH secreted from the adenohypophysis. The spermatozoa are transported by way of the epididymis and vas deferens to, ejaculatory duct from which they are expelled, together with accessory glandular secretions, at the time of mating.

Between the sperm-producing cords within the testes are the interstitial cells of Leydig, which have an endocrine function. Under the influence of the interstitial cell stimulating hormone (ICSH) from the adenohypophysis these cells produce testosterone, the major sex hormone of the male. Testosterone and other male sex hormones are known collectively as androgens. Testosterone causes embryonic development of the male reproductive organs. It also induces the maturation of these or normal function following puberty.

At the time of puberty testosterone also promotes the development of the secondary sex characteristics of males, including physical development, hair distribution, masculine voice, and male behavior. In humans, special muscles called cremasters in the walls of the scrotum elevate the testes close to the body in cold weather or lower them for more effective cooling in warm weather. In all mammals the testes develop first within the abdominal cavity.

In humans they descend into the scrotum, and within a few years after birth the route of descent is obliterated so that they remain permanently there. In seasonally breeding mammals, the testes often return to the intra-abdominal location during non-breeding periods of the year.With the advent of the breeding season, they descend into the scrotum and increase considerably in size.

In few mammals the testes never leave the intra-abdominal site at all.In birds the testes are similarly located in the abdominal cavity, and birds have even higher body temperatures than mammals. Avian body temperature fluctuates quite widely, however, being at its lowest in the early morning hours, in those species that have been studied.

Female reproductive organs

Mammalian ovaries are oblong organs situated in the pelvic portion of the abdominal cavity. Each is close to the expanded opening of a fallopian tube or oviduct which leads centrally to the opening into the uterus. Another opening from the cervix or neck of the uterus leads to the vagina, and this, in turn, opens to the surface of the body.

Each ovary contains a large number of undeveloped sex cells or ova, and as these develop they rise to the surface or the organ and become enclosed in blister-like graafian follicles. At the time of ovulation the follicles rupture and the contained ova, together with some surrounding follicular cells, enter the associated fallopian tube and move through it to the uterus. If mating has occurred and spermatozoa are present, each ovum may be fertilized by the penetration of a sperm cell. The resulting zygote (fertilized ovum) then implants itself into the uterine wall (endometrium) and there undergoes embryonic development. At parturition the developed young are expelled through the vagina.

In humans the maturation of a single ovum in one ovary ordinarily inhibits the simultaneous maturation of other ova so that only a single offspring can result. In humans a new ovum matures every 28 days on the average unless pregnancy occurs.

The human cycle and that of other higher primates differs in still other ways from the cycles of most mammals although all involve a similar interplay of controlling hormones. The human cycle is called the menstrual cycle, derived from the word "menses," or monthly. Starting at puberty and continuing until the menopause at the age of 40 to 50, a new ovum begins to mature in human females every lunar month, Under the influence of the follicle stimulating hormone (FSH) from the adenohypophysis the ovum grows and becomes enclosed in a graafian follicle on the surface of the ovary.

Associated cells of the follicle are endocrine in function and produce a steroid hormone called estrogen. Estrogen has two influences on the adeno-hypophysis: it inhibits the production of FSH and it promotes the production of the luteinizing hormone (LH). As the follicle grows, the secretion of estrogen increases, and as a result FSH levels in the blood progressively fall and LH levels progressively rise. At some critical ratio of FSH to LH, growth of the follicle ceases and it ruptures (ovulation). As the ovum, together with some of the surrounding follicular cells, moves away through the fallopian tube, the site of the ruptured follicle becomes altered in appearance and function. The remaining follicular cells form a new structure called the corpus luteum.

Under the continued influence of LH, the corpus luteum, grows and produces another steroid hormone, progesterone. In humans the corpus luteum also produces estrogen. Progesterone and estrogen both inhibit the production of FSH, so no additional ova can mature in the ovaries at this time. If fertilization and uterine implantation occur, the corpus luteum continues this inhibitory function throughout pregnancy and for some time following parturition. If fertilization does not occur, however, the corpus luteum regresses, FSH production is resumed, and a new cycle begins.

The hormones estrogen and progesterone have a number of functions other than those just described. Estrogen appears during embryonic life and stimulates the development of female reproductive organs, just as testosterone stimulates the development of male reproductive organs. At puberty estrogen and progesterone together initiate the cycle of reproductive activity and continue the development of female secondary sex characteristics. These include the development of mammary glands, typical female fat deposition, hair distribution, treble voice, and female behavior. During the monthly ovarian cycle, estrogen and progesterone cause cyclic changes in the lining mucosa (endometrium) of the uterus.

Under the influence of estrogen during the period of ovum maturation the endometrium thickens. Progesterone from the corpus luteum subsequently causes increased activity of endometrial glands and increased vascularization and sponginess of the tissue. These changes prepare the endometrium to receive and nurture all implanted embryo. If fertilization has not occurred, however, the source of progesterone disappears, the bulk of the endometrium is discarded, and it passes out as the menstrual flow.

This menstruation, as it is called, comes about because the blood supply to the endometrium is discontinued and the endometrial tissues die and are sloughed away. Also, during the monthly cycle estrogen and progesterone cause an increase in the size and development of the mammary glands. If pregnancy occurs, this development continues in preparation for lactation. If pregnancy does not occur, some regression of the breasts takes place until the next cycle begins.

Cyclic changes in behavior also occur and are correlated with the waxing and waning production of estrogen and progesterone. In subprimate mammals no menstrual flow takes Place but the interplay of hormones occurs as in primates. At the time of ovulation the female enters a period of estrus, or "heat", at which time she will receive the male. This occurs under the influence of progesterone. Female rabbits do not ovulate spontaneously but remain in a condition of continuous estrus until nervous reflexes induced by mating cause ovulation to occur.

There are several known estrogens but the most potent naturally occurring one is probably estradiole. It, or a closely related hormone, is present in all female vertebrates. Progesterone, however, is apparently present only in mammals and, perhaps, in birds. Estrogens, as well as androgens, are toxic to cold-blooded vertebrates; consequently, experimental introduction of these hormones has not been a fruitful field of research in lower vertebrates. In very low doses they appear to have effects similar to those in mammals, however. The luteotropic hormone (LTH) was once thought to play a role in the ovarian reproductive cycle, but it now appears that this is true only in rats and mice. In these species LTH augments the effects of LH in the maintenance of the corpus luteum. In other mammals, however, the roles of LTH appear to be confined to the period following reproduction and are concerned, instead, with the care of the young.

When the fertilized ovum has implanted itself into the endometrium of the mammalian uterus, extraembryonic tissues developed by the early embryo cooperate with modifications of maternal tissues to produce the placenta. Besides serving as a membrane across which nutrients, gases, and wastes can be exchanged between the maternal and the fetal circulatory systems, certain cells of the placenta also take on an endocrine function. These cells produce and release chorionic gonadotropins, and in humans estrogen and progesterone as well. The chorionic gonadotropic hormones maintain the corpus luteum, inhibit the menstrual cycle from recurring, and stimulate uterine growth. Chorionic gonadotropin in the urine forms the basis for pregnancy testing since other vertebrates react in predictable ways to injections of this hormone or urine samples containing it. During pregnancy the secretions of a number of other endocrine glands are also increased. Among them are those secreted by the thyroid, parathyroid, and adrenal cortex.

The large quantities of progesterone present during pregnancy inhibit uterine muscular contraction, but shortly before the time of parturition (childbirth) increases in estrogen secretion overcome this inhibitory effect. The uterus then becomes responsive to the stimulatory effect of oxytocin from the neurohypophysis. When it becomes sufficiently sensitive to oxytocin, rhythmic contractions of the uterus begin. These lead to expulsion of the child and soon thereafter of the placenta as well. This abrupt reduction in chorionic gonadotropin, estrogen, and progesterone has major effects upon the mother and sometimes upon the child as well. The developing child has been exposed to the same hormonal environment as the mother.

The mammary glands of both sexes, have been stimulated, and it is not unusual for some small amount of lactation to occur from the newborn as we'll as from the mother. For similar reasons, newborn female children sometimes show some small menstrual flow for a day or so after parturition. The role of the luteotropic hormone (LTH), sometimes called prolactin, in mammary response is currently in doubt. It was once believed that this hormone initiated milk secretion in a mammary gland which had been prepared anatomically by the combined action of estrogen and progesterone.

More recent research has shown that the growth hormone (STH) functions just as effectively as LTH and that the two hormones may be identical in mammals. In any case, following parturition oxytocin causes ejection of milk when the nipple is physically stimulated by the nursing child. Another hormone, relaxin, has also been shown to play a role in parturition. This hormone is produced by the ovary, placenta, and uterus and -it may have some unknown function during pregnancy, since the blood levels of relaxin progressively increase during this period. At the time of parturition, relaxin is responsible for the dilation of the uterine, cervix and the vagina.

Chapter 5

Evolutionary Biology

Charls Darwin is proclaimed as the father of today's evolutionary theories. Darwin's thinking about the mechanism of evolution was greatly influenced by his observations on selection of domestic animals and plants by man. Through consistently continued selection of specific individuals for breeding man has produced many different varieties. Darwin was particularly interested in the many breeds of pigeons that had been produced from what was originally one species. He might just as well have been interested in the many breeds of dogs. Some of these have been produced within relatively recent years, and the history of their production is known. By artificially selecting parents that have particular characteristics a breeder can gradually produce desired body patterns. Similarly man has been able to increase the production of milk by cows, the yield of grain by corn, and the speed of horses.

Darwin, of course, did not suggest that some human being was doing this in the wild animal world. But he did ask whether there could be a possible means by which selection for particular characteristics could go on in nature—in short, was there any mechanism for natural selection? His studies soon revealed that indeed there was such a mechanism, enabling him to propose a theory each step of which was susceptible to testing.

All animals and plants produce more offspring than are needed to replace themselves if all the offspring were to live. In some cases this overproduction is enormous. Even slowly reproducing animals, such as elephants or man, produce many more offspring than are necessary to replace the parents. This fact is becoming uncomfortably familiar to all who are concerned with the human population explosion. This has resulted from the fact that disease and famine which formerly killed children have been largely overcome, and man's rate of reproduction is now far greater than what is needed for mere replacement.

In spite of the fact that animals and plants overproduce offspring, the population of a species in a given region over a given period of time remains relatively constant. There are variations from time to time, and sometimes populations may rise or drop, but on the average the number of individuals of a species remains relatively constant. After a hundred of years of study, there is no reason to doubt the truth of these first two points.

Competition for survival and reproduction from, these we logically infer that there is competition for space, food, and mates. It is obvious that there is not enough food nor is there enough space for all the individuals produced by a single pair of animals to continue to reproduce. In the foregoing housefly example, for instance, the crowding that is postulated could not practically occur. The flies could find neither enough food to live nor enough space to move about.

Thus, there is a constant struggle within the species and with other species for space, food, and all the other needs for life. The ultimate competition is competition to reproduce, to pass on life to the next generation. This postulate, while not as directly observable as the first two, is a logical deduction from these, and where tested experimentally has been found to be true.

In any population of animals no two individuals are alike. This observation is a truism when we think of the animal that we know best, namely man. Those of us who have had pets know that it is also true with respect to dogs, cats, or birds. It may not seem to be true with animals such as insects, but when one comes to study these carefully he finds that they too vary among each other, so that no two are alike. These variations mean that animals face competition, face the environment, with different chances for survival and reproduction. If variations among individuals are to have any influence on succeeding generations, they must be heritable—that is, they must be capable of being passed on to the offspring. It was on this point that Darwin ran out of knowledge, so to speak.

Those individuals that are best adapted to survive do survive to pass on their variations to future generations, and those that are poorly adapted do not. By this means, adaptive characters increase in proportion in the population, and characters that are not adaptive decrease, thus producing overall adaptation. This is the crux of natural selection. The environment in its totality, in a sense, selects which of the many available variations is to be passed on to succeeding generations. However, in time, this would produce a stable situation, as it became increasingly more difficult to grow more adapted. Thus, one might expect that gradually perfectly adapted species would arise, and evolution would cease. Such is not the case, however, because of one other fact.

Changes constantly occur, not only short-term changes such as severe storms, fires, and floods, but also long-term changes such as elevation of mountains and rising of seas. Therefore, large groups of animals may at one time be remarkably well adapted to live in particular environments, as were the dinosaurs, but at another time, with new environments, may be unable to live at all. The occurrence of evolution is now regarded by biologists as almost infinitely probable. There is little or no evidence that gradual change with time, descent with modification, has not been the rule. Dartoinism—that is, the explanatory theory here described—as modified by newer research, is considered as highly probable, the best explanation that we have so far.

First, we shall show how the idea of evolution helps us to understand the means by which animals have solved the many problems that face them in staying alive. We shall do this through a study of structures and processes used by animals to solve their life problems. Then we shall cover the discoveries of genetics about variations and their inheritance, to see what this teaches us about the spread and loss of variations through populations. Following this, we shall cover what modern ecology reveals about the environment and its influence on animals and plants. Armed with this information, we shall then attempt a modem synthesis of ideas about the mechanisms by which evolution occurs.

Although past endeavours are regarded by many as irrelevant to modem biology, no apology is made for beginning with a historical overview. The topic at hand has a long and involved history, and changes in organisms over time are its very foundation. That said, one enters this arena with trepidation. One major difficulty in looking for, relationships between development and evolution is that we are dealing with what have been regarded as two self-contained biologies:

i) Functional/physiological causation, which deals with all aspects of the activation and operation of genetic programmes within an individual lifetime; and

ii) Evolutionary causation, viewed traditionally as dealing with how genetic programmes are altered through evolutionary time so that new variants arise.

Physical embodiment of proximate versus ultimate causation, in some animals, is in the separation of their eggs into soma and germ plasm. In such animals the somatoplasm, from which body cells develop, provides the information for the individual organism during its lifetime. The germ plasm is inherited by germ cells that produce gametes.

Genes within these germ cells, together with gene products inherited maternally, provide the heritable information for subsequent generations and the raw material for future evolutionary change. Importantly, germ plasm is not segregated from the soma in most animals or in any plants or fungi. Somatic inheritance is a theoretical possibility in such organisms and modes of inheritance therefore an important topic. The two self-contained biologies are not so self-contained in these many organisms. Part of the connection between evolution and development is etymological.

Evolution as change of gene frequencies

Population genetics, the branch of genetics that deals mathematically with frequencies and interactions of genes in populations, is especially concerned with how such factors as mutation, selection and migration alter gene frequency. Evolution as change of gene frequencies in populations was the most widely accepted definition reflecting the domination of evolution by population genetics. This domination is now over. Not that population genetics is an inappropriate means to study evolution—a quantitative genetics model of morphological changes in evolution. Rather it is neither sufficient nor inclusive. The search for the basis of the adaptiveness of small—and large scale gene effects by Palopoli and Patel reveals that identification of genes does not necessarily reveal the developmental path ways underlying evolutionary change. Evolution is hierarchical. Although population genetics is not often used to shed light on macroevolution carefully designed studies show how such an approach can illuminate the evolutionary history of a group.

The analysis by Futuyma, Keese and Funk of an ecologically important character—host affiliation—in the leaf beetle *Ophraella* is such a study in which limitation on genetic variation influences the evolution of host affiliation, i.e. macroevolution is limited by a genetic constraint. Although all evolutionary change is based on changes in genes, evolutionary change can occur at the organismal level without being reflected at the supraorganismal level. Nor need change at the supraorganismal level be qualitatively different from change within individuals.

A specific microevolutionary explanation for an apparent macroevolutionary transformation—the origin of external from internal cheek pouches in mammals. Because evolution acts at all these levels, no single, universally applicable definition exists even today for change over time. There is much to be said for Darwin's 'descent with modification' as a definition of evolution. General enough to encompass all levels, it emphasize; the temporal, genealogical and variational aspects of evolution.

Development is also hierarchical, but over the time-scale of individual ontogeny. Developments are those complex hierarchical processes within an individual that produce and order cells, tissues and organs through the processes of differentiation, morphogenesis, and growth. Not all evolutionary change is mediated by changes in development. Aspects of organismal evolution that are mediated by development include development itself, embryonic stages and processes, morphology and phenotypic plasticity. Patterson provided a particularly useful discussion of the relationship of ontogeny to phylogeny, emphasizing that most evolutionary laws also describe ontogeny, pertaining as they do to growth, size and specialization of individuals.

Thus, Dollo's law of the irreversibility of evolution concerns the inability of an organism to return to a state identical to an ancestral condition; Cope's law of the unspecialized states that evolutionary novelties in new taxa are more likely to arise from generalized than from specialized ancestral taxa.

Cope's rule of the tendency of animals to increase in size during evolution carries connotations of progress, trends and increase in complexity, while Williston's rule is that there is a decrease in the number of serially repeated parts as individual parts specialize during evolution. Dollo's law is discussed to illustrate how, at the level of developmental mechanisms, transformations in ontogeny produce phylogenetic change. Because such laws and rules deal with development as much as with evolution, Patterson argued that phylogeny hardly differs from ontogeny but saw ontogeny as economical and phylogeny as a 'random walk'.

Atavisms, phylogenetic character reversals, and the reactivation of silenced genes, all indicate how knowledge of developmental processes reveals the ontogenetic basis of Dollo's law. Although ontogeny is individual development and phylogeny evolutionary relationships, once we begin to consider mechanisms of change through time in different organisms, the two are integrated as evolutionary developmental biology.

The challenge taken up by evolutionary developmental biologists is to integrate development with genomic, organismal and population approaches to evolution. Evolutionary developmental biology strives to forge a unification of genomic, developmental, organismal, population and natural selection approaches to evolutionary change. All these elements have been brought to bear on how phyla and the body plans that typify phyla arose.

Evidence of evolution in vertebrates

Acquisition of new structures or functions could come about through correlated changes in developing systems, pleiotropy, epigenetic interactions and/or intensification or alteration in function. Most evolutionary changes in morphology are probably of these types. Change must, of course be associated with fitness, and both rate and type of change depend on the genetic variance in fitness and/or on traits that are correlated with fitness. Components of functional complexes may be integrated by additive or synergistic interactions, as Sharon Emerson and her colleagues demonstrated so elegantly for integration of the morphological characters accounting for 'flying' in frogs. The functional stimulus for such changes could involve intensification of function or acquisition of a new function. Because functionally interdependent structures are tightly coupled epigenetically in developments, innovations could arise as secondary by-products of epigenetic change during development, or as side-effects of phylogenetic change in size or developmental timing.

Bolitoglossine salamanders have one of the most, if not the most, specialized feeding systems of any amphibian. The tongue, used as a projectile, is 'fired' rapidly and with enormous accuracy at the prey, over distances as great as 80% of the length of the body. Depth, distance perception and a highly specialized musculoskeletal feeding apparatus are required for such specialized and effective feeding, which is so accurate that these animals virtually never miss their prey.

At the extreme of this specialization is the genus *Hydromantes* from California, Italy and France in which, not only the tongue, but portion of the visceral skeleton is projected outside the mouth using a ballistic mechanism. Changes in feeding in bolitoglossine salamanders from the ancestral condition are associated with the acquisition of direct development, although loss of the larval stage is a necessary but not a Sufficient condition.

David Wake and Gerhard Roth have documented evidence from the functional morphology of feeding behaviour in plethodontid and bolitoglossine salamanders for what they term 'ontogenetic repatterning', by which they mean the establishment of new sets of morphogenetic processes resulting from heterochronic events. Ontogenetic repatterning is not so broad a concept as to embrace any change in development or any addition, deletion or compression of an existing or ancestral ontogeny. Ontogenetic repatterning establishes a new developmental trajectory leading to change that is structurally and functionally integrated.

In their 1985 analysis Roth and Wake identified the following repatterning events in modification of bolitoglossine feeding mechanisms:

i) Suppression of branchial arch development so that only a single epibranchial element develops fully;

ii) Extensive elaboration of the tongue musculature required to project the tongue so far and fast;

iii) Rotation of the eyes to the front of the head from the more lateral position typical of most other amphibians; and

iv) Development of ipsilateral retinotectal projections for stereoscopic vision, comparable to those seen in mammals.

Ontogenetic repatterning provides a powerful developmental process upon which selection can act in the introduction of new adaptive innovations. Fran Irish proposed a shift in timing of development and resulting paedomorphosis associated with the capture of large prey items as instrumental in the evolution of the snake skull from the more generalized skull of their lizard-like ancestors. The changes included reduction of the chondrocranium, elaboration and enhanced ossification of the dermocranium, a mobile joint between the mandibles, and reduced connections between bones of the snout, palate and jaws.

Deleting these connections facilitated the increased mobility of the snake skull and jaws so essential for capturing and swallowing large prey items. A functional/ morphological approach involves investigation of the integration of changes in many systems and, consequently, extensive knowledge of such diverse fields as morphology, development, physiology, neurobiology, biomechanics, ecology, systematics and evolution. These tasks are daunting but necessary if we are to uncover mechanisms for the initiation of morphological changes. Studies on pharyngeal jaws are a model. Such studies move us away from identification of single key innovations and toward an emphasis on integrated changes and ontogenetic repatterning in interrelated.

Examples of evolutionary change resulting from intensification of an existing function include long-term improvement in the mechanical efficiency of the skeleton leading to the evolution of the artiodactyls; highly modified digging claws and teeth in taeniodont mammals; evolution of the hypocone on mammalian teeth; and such neomorphic features as the divided maxilla of bolyerine snakes and secondary jaw articulations in birds. Primitive mammalian upper molar teeth are triangular with a cusp at each of the apices of the triangle. The cusps are known as the paracone, metacone and protocone. On at least 20 independent occasions in mammalian lineages since the Caenozoic a fourth cusp has evolved. Situated at the disto-lingual corner of the upper molar teeth, the hypocone doubles the surface area available for crushing food.

Hypocones evolved frequently and are only rarely lost, attesting to the advantages associated with converting a three-cusped to a four-cusped molar. Jernvall and Hunter and Jernvall consider that the hypocone fits Mayr's definition of a key innovation precisely: a new structure allowing a new function that opens up a new adaptive zone to the organism. They document various ways in which the hypocone may be produced and argue for the presence of intermediate conditions in extant and fossil mammals either an enlarged shelf or a small cone not in occlusal contact with the lower molars. They further document that hypocones evolved in a great diversity of mammals on all the major continents; that Caenozoic mammals with hypocones underwent much greater radiations than did forms without hypocones; and that hypocones are preferentially found in herbivorous mammals.

The hypocone allows invasion and diversification within herbivorous adaptive zones. As a high proportion of generalist feeders possess hypocones, Jernvall and Hunter suggest that hypocones evolve under generalist dietary regimes and are then pre-adapted for development of the more specialized herbivorous diet requiring additional tooth area to crush and grind food.

Bolyerine snakes of the species *Bolyeria multicarinata* and *Ceasarea dussumieri* are a subfamily of boid snakes found only on Round Island, near Mauritius in the Indian Ocean. As with all living boid snakes, they possess vestigial hindlimb skeletons and/or traces of pelvic girdles. Their maxillae are divided into anterior and posterior articulating halves; this condition, unique among vertebrates, probably results from failure of fusion of the two centres of ossification from which the maxillae form. The development of secondary jaw articulations in birds is a particularly nice example of intensification of function driving an evolutionary novelty.

With the strong musculature required for the intensified impacts associated with feeding in many birds came the need to increase the surface area of the jaw and skull skeleton for muscle insertion. This was met by the development of a bony projection from the mandible which grew medially toward the skull. If the bony process comes in contact with the skull, the differentiative properties of the osteogenic cells of the bony process facilitate the development of an articulating surface between the bony process and the skull. Contact between these two previously separate parts of the craniofacial skeleton is sufficient for a new articulation to develop, giving these birds two jaw articulations—the original, primary quadrate-squamosal articulation and the new, secondary mandible-skull articulation.

As in adaptations associated with pharyngeal jaws in cichlid fishes, the adaptive potential of periosteal cells responds to functional demand. Walter Bock estimated that two forms of secondary articulation—between the mandible and the basitemporal and between the mandible and the ectethmoid—evolved independently in 15 and three avian lineages respectively.

The evolution of these secondary articulations was achieved readily and rapidly through simple shifts in ontogeny, a mechanism which Bock termed multiple evolutionary pathways. If intensification of function initiates their development, then the presence or degree of development of secondary jaw joints should be correlated with intensity of the impact of feeding or prey capture on the mandible, just as the primary jaw articulation reflects intensity of functional demands. This is indeed the case. In the plover *Charadrius*, where impact is low, the joint is a syndesmosis of fibres and ligaments without articular cartilage or cavity.

The marine skimmer, *Rhynchops nigra* on the other hand, flies just above the waves, dipping its knifelike lower beak into the water to catch fish. The head can be thrown back with sufficient force to break the horny tip on the bill. It will not be a surprise to learn that *Rhynchops* has the most robust and highly developed secondary jaw articulation consisting of a true diarthrodial joint with articular cavity, articular membrane and dense cartilaginous articular surfaces. Such changes are integrated. Intensification of function, developmental and functional association of skeletal and muscular systems, differential growth, and the cellular capabilities of osteogenic cells for joint formation, combine to facilitate formation of a new structure whose function is just sufficiently different from that of the original jaw joint that both persist.

Examples of evolutionary change associated with alteration of function or acquisition of a new function include:

i) Transformation of the reptilian to the mammalian jaw articulation, and the development and integration of ossicles into the mammalian middle ear;

ii) Transformation of the first dorsal fin ray into a lure in angler fish;

iii) The origin of the thyroid gland from the endostyle;

iv) Development of the turtle shell from ribs and vertebrae;

v) Derivation of teeth from scales in various groups of fishes; and

vi) origin of the snake skull from a lizard-like skull.

Transformation of the reptilian to the mammalian jaw articulation, and the concomitant freeing of bones for inclusion as ossicles in the mammalian middle ear, is a resounding example of triumphant morphological analysis.

Following Kemp, who used the term specifically for the changes associated with the transition from reptiles to mammals, Thomson termed such transformations and integrated changes in interrelated systems in the ear as 'correlated progression'. Mammals have only the dentary as a single bone in the mandible. Reptiles have at least seven bones—dentary, articular, prearticular, surangular, angular, splenial and coronoid. Articulation of reptilian jaws is effected through the articular of the lower jaw and the quadrate of the upper jaw.

In mammals, articulation is between the dentary of the lower jaw and the squamosal of the skull. Marsupials represent a fascinating intermediate condition. Neonates of the South American possum *Monodelphis domestica* have neither the mammalian dentary-squamosal nor the reptilian quadrate-articular jaw articulations.

Acquisition of the mammalian condition can be followed in marsupial embryonic development and also through an excellent fossil record of the reptile → mammal-like reptile → mammal transitions. During evolution of the mammal-like reptiles, the dentary expanded at the expense of the other bones of the lower jaw. With expansion, the dentary came into contact with the wall of the skull, and as happened in birds and cichlid fishes, periosteal cellular reactivity at the contact site initiated the formation of a joint.

With the development of this new squamoso-mandibular joint, functional control was removed from the articular-quadrate joint, freeing these two bones to take on new functions.The articular and quadrate gradually moved posteriorly away from the original jaw articulation. This condition is seen in fossil synapsid reptiles. Ear ossicles, which were present in synapsid reptiles, were co-opted by the developing mammalian middle ear to form the malleus and incus.

The mammalian hearing system not been available for selection at the same time as the transformation in jaw articulation was taking place, the articular and quadrate would in all likelihood have become vestiges or disappeared in mammals as have the other reptilian jaw bones. They would not have been retained as key players in this adaptive scenario. Even though the sequence is known in exquisite detail, it is still not clear whether the onset of the mammalian condition was driven by a functional demand for high frequency hearing, by developmental changes in the jaws and jaw musculature, or by differential growth between brain and jaw.

Dolls Law—the law is named after the Belgian palaeontologist Louis Dollo by one of his students—is concerned with reversibility of evolution. Dollo maintained that organs, other complex structures and indeed whole organisms, cannot return to a condition identical to an ancestral condition.

Dollo's law appealed particularly to palaeontologists, whose discipline relies so completely on morphology and interpretation of apparent evolutionary trends in morphology. The reappearance of structures lost in ancestors (such as the second lower molar in fields), atavistic structures (such as limbs in snakes, hindlimb skeletons in whales and additional digits in horses.), or phylogenetic character reversals, affirm that toss of a structure need not necessarily mean loss of the ability to form the structure.

Dollo's law thus has substantial embryological implications: traces of developmental stages can be retained in descendants, especially as ancestral structures in early ontogeny. Dollo's law was tested by Marshall, Raff and Raff in the light of current knowledge of rates of degradation of genetic information. The premise upon which the test is based is that evolution could not be reversible if genes or developmental programmes rapidly become non-functional after they have been released from the selective pressure associated with maintenance of the organ or structure. If information is lost rapidly (as is the conventional wisdom), a structure could not reappear exactly as it had been in an ancestor. They determined, however, that there was a significant chance that silenced genes or apparently lost developmental programmes could be reactivated if the time scale of loss was of the order of 0.5-6mya, but not if loss were more than 10mya. If selection maintains the genes or programme for another function, then even long-silenced genes or programmes have the potential for reactivation.

A silenced gene is defined as a gene that is no longer transcribed; i.e. the gene might still be present but inactive, as indeed it must be for an ancestral pattern to be reestablished. It will therefore be important to gather similar data for regulatory gene products, as emphasized by Zuckerkandl in his approach to parallel evolution via modification of common molecular pathways. Atavisms are ancestral characters that occasionally reappear in individuals of descendant species. This is reflected in synonyms for the word, which include 'throwbacks', 'reversions', and 'reproduction of the ancestral type'.

Atavisms are traditionally interpreted phylogenetically as the reappearance of an ancestral state.

When viewed ontogenetically they provide evidence of the retention of ancestral developmental programmes. Situations in which a 'lost' structure 'reappears' tell us a great deal, not only about how the structure was lost, but also which portions of the developmental programme were lost, rather than silenced or modified.

Indeed, such atavistic reappearances tell us that loss of tile structure cannot have meant loss of the developmental machinery to produce the structure. The genetic basis of atavistic characters was established in a large series of breeding experiments on guinea pigs with extra toes, undertaken by Stockard, Castle, Sewall Wright and Scott in the first three decades of this century. Surprisingly, structures are often integrated with adjacent tissues and organs as functional components of the organism. Perhaps the classic examples from nature are three-toed horses, hindlimb skeletons in whales, atavistic muscles in passerine birds and prolegs in insects. Other examples, not discussed here, include atavistic muscles in mammals, winged earwigs, molar teeth in Pleistocene *Lynx*, extra nipples, tails and excess hair in humans, dew claws in clogs, an ascending process of the astragalus in bird embryos, a vestigial ego, tooth in some marsupials, and patterns of bristles in *Drosophila* that are diagnostic for flies in a related family. Atavisms in basal taxa can point the way to structures that will appear in more derived taxa, as well as telling us what has gone before.

Evolution of modern horses of the genus *Equus* involved numerous changes, including reduction in the number of toes from four in the forelimb and three in the hind of *Eohippus* (Early Eocene) to a single elongated toe in fore- and hindlimbs of *Pliohippus* in the Pliocene and *Equus*. Elongation of the single toe accompanied reduction in the length of humerus and femur and even greater reductions in ulna and fibula, including loss of their distal articulating surfaces. Such extreme reduction in lateral digits was not unique to the horse lineage.

Ungulates from South America (Order Litopterna) exhibits similar changes. Embryonic horses have primordia for three toes of equal size and development. During ontogeny the middle toe increases its rate of growth and development, leaving the two lateral toes behind until they are no more than vestiges (splint bones) lateral to the cannon bones (fused metacarpals or metatarsals). But if a mutation allows the lateral toes to continue to grow they can develop as fully formed digits no smaller than the middle toe. Julius Caesar and Alexander the Great prized their war horses with extra toes. The mustangs of the western United States are said to possess extra toes more often than do other breeds. The ability to modulate the growth rate of the lateral toes leads to subsequent and higher order interactions between skeleton, muscles, ligaments, tendons, and the nervous and vascular systems. Extra toes are in contact with the ground and integrated functionally into the musculo-skeletal system.

Whales in the Order Cetacea are all marine but evolved from terrestrial ancestors in the Paleocene or Eocene epochs. Limbed and terrestrial ancestors of modern-day aquatic whales could have ventured into water before flippers evolved, just as aquatic amphibians ventured onto land before tetrapod limbs had evolved fully. All living whales lack hindlimbs and have forelimbs modified as flippers. Before forelimbs were modified as flippers and hindlimbs were lost, aquatic locomotion in whales was probably powered by undulation of the vertebral column rather than by the limbs.

As in tetrapods, turtles and so many other taxa, the features of modern whales appeared sequentially and not as an integrated package. The sperm whale (*Physeter catadon*) is the only toothed whale to have a vestigial hindlimb skeleton, although expression is variable, ranging from a single pelvic bone to presence of a nubbin of a femur, rudimentary fibia, fibula, tarsals and metatarsals.

Although living baleen whales lack complete hindlimb skeletons as adults they do retain rodlike vestigial pelvic bones or cartilages, femora and occasionally tibiae that fail to project beyond the body wall. Atavistic hindlimb elements can be surprisingly complete, for example, 12.7cm long left and right hindlimb elements in a female humpback whale; 15cm diameter elements in a female sperm whale.

Passerine birds are recognized by a variety of characters including absence of two leg muscles, *Musculus iliofemoralis externus* and *M. caudiliofemoralis pars iliofemoralis*. *M. iliofemoralis externus* is readily recognizable from its origin on the iliac crest and insertion on the femur. It is found as a normal element in ducks, falcons and galliform birds but is absent from kingfishers, hornbills, trogons, toucans, woodpeckers and cuckoos, which are phylogenetically closer to the passerines.

Gradual improvement in function and availability of new habitats are just as likely initiators of change as key innovations. Often change in structure/function and habitat proceed hand-in-hand, as in the proposed origin of book-lungs from book-gills in the transition from aquatic to terrestrial scorpions in the Carboniferous, insect wings from limbs or gills, limbs from fins, or feathers from scales. Invertebrate examples include the evolution of wings in insects, torsion in gastropods and eyespot patterns in butterfly wings.

For vertebrates, examples include the origin of jaws and transformation of jawless to jawed vertebrates, the transition from fish fins to tetrapod limbs, the origins of birds and feathers, evolution of turtle shells, and the development of pharyngeal jaws in cichlid fishes. The earliest known vertebrates of the Cambrian Era are jawless fishes possessing a series of gill arches, each with a skeletal component supporting the respiratory organs. The fossil record of these jawless fishes is very complex and subject to several interpretations.

In the traditional scenario of the transition from jawless to jawed vertebrates, the most anterior of these gill arches was extended anteriorly to a position supporting the mouth and buccal cavity, carrying its cartilaginous supporting element with it. The dorsal portion of the arch formed the palatoquadrate, the skeleton of the upper jaw; the ventral portion formed the mandible, the skeleton of the lower jaw. The necessary elements were thus present in the original gill arch and functioned in respiration, but were co-opted for a new function (feeding) in this new position. No new gill arch had to be constructed, no new' skeletal elements had to evolve.

The traditional view is that enhanced feeding was the driving selective force, but whether feeding, predation, or respiration drove this evolutionary change is uncertain. Evidence that jaws are modified gill arches comes in part from development, in part from homology of the skeletal elements, and in part from the equivalent developmental origins and inductive interactions responsible for their formation.

Like the jaws of gnathostomes, the velum is derived from the mandibular arch and may be the agnathan homologue of the jaws. These alternative scenarios emphasize how critical it is to establish homologies of the structures and organisms being considered, before proposing an evolutionary scenario. The key innovation in the agnathangnathostome transition could be regarded as behavioural, functional /ecological, structural, developmental or size-related, depending more on individual inclination than on evidence for the change itself. Arguments for each of the four areas are:

i) *Functional/ecological*: Increasing toughness of potential food and the need to develop a more robust food-gathering apparatus;

ii) *Structural*: Availability of the first visceral arch with its skeletal elements and associated musculature;

iii) *Developmental*: The transformational capabilities of the mesenchymal and skeletal tissues of the arch; and

iv) *Growth-related*: The capability of the arch and its components to grow anteriorly.

To isolate one as the key innovation is certainly misleading and other innovations were taking place in these same organisms. However they arose, jaws, once in place, provided a key innovation for the 500-million-year radiation into every conceivable and even some inconceivable habitats, allowing jawed vertebrates to capture and utilize an extraordinary range of prey and foodstuffs. Comparative studies of structure and development can be augmented by analyses of sets of genes or even of individual genes.

With the rapid growth in comparative molecular biology over the past seven to ten years has come knowledge of molecules for which sequences and partial sequences are now known for a sufficient number of organisms that they can be used for broad-based phylogenetic analyses and/or for analysis of key events accompanying or accounting for the origin of individual groups of organisms. Sidow used 55 partial sequences of the ten *Wnt* genes then known in an investigation of the initial diversification of jawed vertebrates.

Wnt-1 to *Wnt-10* are ancient, having diversified before the last common ancestor of arthropods and deuterostomes some 600 or more mya. A round of duplication of *Wnt-3*, -5, -7 and -10 occurred early in the lineage leading to craniates. Subsequent amino acid replacement in *Wnt* genes was four times lower in jawed vertebrates than in agnathans and echinoderms. The implication is that *Wnt* genes were serving an important function(s) regulating development in the 100 or so million years associated with the origin of jawed vertebrates.

A further implication is that *Wnt* plays or played more important roles in the development of jawed than jawless vertebrates, although the hagfish Eptatretus stoutii has several *Wnt* genes. Sidow generalizes from the expression of *Wnt* in limb buds, the neural crest, and central nervous system, to argue that diversification of *Wnt* genes facilitated generation of cell types and developmental pathways characteristic of vertebrates.

An excellent example illustrating the utility and potential pitfalls of analysing homology on the basis of patterns of gene expression, is expression of the segment Polarity gene engrailed in the lamprey *Lampetra Japonica*. As in other vertebrates, *engrailed* is expressed at the midbrain-hindbrain boundary and in mesoderm of lamprey mandibular arches.

At this level of analysis the expression patterns suggest conservation of engrailed across the vertebrates. Later its development, expression of *engrailed* is restricted to one particular set of muscles, the velothyroideus muscles that activate the velum. Homology of the lamprey velum, is unclear. Two of the jaw muscles in zebrafish, the levator arcus palatini and dilator operculi, express *engrailed* at an early condensation stage in their development. While accepting the difficulties of using expression patterns of isolated genes to establish homologies of the structures expressing those genes, Holland and his colleagues favour the view that the velothyroideus muscles in the lamprey are homologues of the two *engrailed* positive muscles in zebrafish, and by implication, in all teleosts.

The inference—that the lamprey velum is the homologue of the gnathostome jaw and that jaws arose from the velum—illustrates the dilemma. Are we justified in inferring the origin of vertebrate jaws on the basis of the expression pattern of one gene? Most of us would answer with a resounding no! Even a comparative analysis of engrailed expression across the teleosts, while it would reveal the generality of the pattern, would not resolve the fundamental difficult of whether the pattern of expression of a single gene is an adequate basis to establish structural homology and/or to infer evolutionary origins. Even analysis of patterns of diversification of many, genes might not be instructive, given that multiple genes would change when a new pathway was co-opted. Patterns of gene in conjunction with more traditional measures of homology are much more likely to reveal patterns of origination.

Determination of engrailed expression in parallel with analyses of homology based on muscle condensations and development in primitive fishes may provide a base from which to tackle such fundamental problems as the origin of jawed vertebrates. Combined molecular, developmental and phylogenetic analyses of such developmental regulatory genes hold great promise. Paired lateral projections which were not appendages, and which consisted of spines and plates as modifications of dermal armour, first appeared in vertebrates in

Silurian and Devonian jawless ostracoderms. Such exoskeletal projections are not homologous with paired fins and limbs: they were ectodermal and not mesodermal, stabilizers rather than locomotory organs. Flexible anterior and/or posterior sets of fins are first found in various Devonian fishes.

Just as median impaired dorsal and ventral fins preceded paired fins, so pectoral preceded pelvic fins. Structural evidence, essentially based on position and connections as proposed by, Geoffroy, makes the case for the transition from fins to limbs, pectoral and pelvic fins transformed into fore-and hindlimbs when a group of Devonian lobe-finned fishes ventured onto land. As Michael Coates and Jennifer Clack have documented from the new fossil evidence, this major ecological transition from water to land was neither accompanied by, nor perhaps even facilitated by, a simultaneous transformation of the fins. *Panderichthys* and related sarcopterygian fishes from the Devonian appear to lie closest to the first tetrapods. *Ichthyostega* and *Acanthostega*, two of the first tetrapod genera, were partly aquatic. Their pectoral fills, like those of earlier sarcopterygian fishes, are composed of a single proximal element and two more distal elements regarded as homologous with the humerus, radius and ulna of the tetrapod limb. Elements equivalent to wrist/ankle elements are present in *Ichthyoslega* and Acanthostega, as they were to varying extents in earlier sarcopterygians.

Although elements equivalent to proximal phalanges may be present, no distal elements equivalent to the digits of the tetrapod limb are present. They appear to be a tetrapod invention. According to one Iona-standing view, lateral fin folds set the stage for the evolution of fins and then tetrapod limbs. According to this theory, known from its original proponents—is the Thacher-Balfour or Thacher-Mivart-Balfour fin-fold theory of the origin of the paired fins, paired fins share common elements with the median unpaired fins in that both are derived from folds of the body wall and supported by skeletal elements. As demonstrated from the fossil evidence, however, pectoral and pelvic fins did not evolve simultaneously among primitive fishes. It is unclear whether this necessarily argues against their origin in a continuous fin fold; the separation need not have been contemporaneous. That no extant vertebrates possess such a continuous lateral fin fold that can be related to orthodox fins, may be a more compelling reason against the fin-fold theory. A complex of interactions between mesenchyme and ectoderm regulates proliferation, outgrowth and patterning along the three limb axes. Fins also begin development as a core of mesenchyme surmounted by an ectodermal thickening.

Most extensively studied in the trout, this 'pseudoAER' or fin fold does not remain as a prominent projection but rather folds onto itself and increases in Dermal fin rays, which make up the distal fin elements in fishes, develop within these ectodermal folds. Mesenchyme of the fin rays is of neural crest origin. It is presumed, but not yet proven, that the lepidotrichia and actinotrichia that make up the fin rays themselves are also of neural crest origin. With respect to the tissues that form them, the dermal exoskeleton is based on bone and associated dentine (±enamel).

The endoskeleton (which may be of neural crest or mesodermal origin) is based in cartilage which may either persist as a permanent cartilage, or be replaced by being surrounded by (perichondrial) bone or invaded by (endochondral) bone. The proximal fin endoskeleton of teleosts, elasmobranchs and chondrichthyes is mesodermal. In teleosts, bony fin rays are the major component of the fins; the proximal endoskeletal is a minor component. The presence of neural crest derivatives in fins is a fundamental difference from tetrapod limbs, which lack any neural crest (dermal) elements, and in which the endodermal skeleton has expanded to include formation of digits.

From such descriptive analyses comes the surmise that the transition from fins to limbs involved:

i) Elimination of the dermal fin rays;

ii) Elaboration of the existing proximal endoskeleton;

iii) Elaboration of the distal endoskeleton for wrist, ankle and the proximal phalanges and

iv) Formation of new distal endoskeletal elements as digits.

These changes are essentially the loss of the neural crest component and a take-over by the mesoderm-derived endoskeleton. In 1991 Peter Thorogood proposed, and in 1995 Sordino and colleagues elaborated, model of altered timing to explain these shifts in the fin $\rightarrow$ limb transition. Thorogood related development of the endoskeleton to the degree and timing of folding of the ectodermal fold. Alteration in timing of appearance of the apical fin fold would prolong the time during which mesenchyme was in association with the ectodermal proliferative signal. The consequent extension of signalling from ectoderm to distal mesenchyme would promote formation of additional distal mesenchyme.

Such a timing switch or heterochrony, when it changes an evolutionary pattern, would explain how the autopod of the limb bud was extended in the fin-limb transition, but does not explain how that mesenchyme came to form novel limb structures such as digits. We still do not know how fins acquired the distal chondrogenic condensations necessary to produce the tetrapod novelty of an autopod with the original sarcopterygian carpals and/or tarsals, to which digits were added.

Molecular evidence from limb development contributes to our understanding of this aspect of the fin → limb transition. *Hox* genes position limbs along the body axis and play regulatory roles in limb bud development. Similarities in cellular development of fins and limbs have prompted investigators to ask whether *Hox* genes play the same role in fin as in limb development. We now know that they do. Can transitions in a Hox code between fins and limbs explain the formation of one from the other, especially the formation of the skeleton of the new digits? Perhaps yes, perhaps no.

It certainly appears, at least from knowledge of zebrafish development, that unpaired median fins do not express the Hox genes (*Hoxa* and *Hoxd*) that pattern paired appendages. If this is the ancestral condition, then Hox genes had to be co-opted to pattern paired appendages.Similar expression boundaries of Hox and homeotic genes are found in fin and limb buds. Admitedly not many species of fish have been sampled; for that matter, not many species of tetrapods have been studied either. Nevertheless, the similarities are very close. Some of them are:

i) Expression patterns of *Hoxc-6* and the *Hoxd* and *Hoxa* clusters in the fin buds of the zebrafish *Danio rerio* are equivalent to those seen in the limb buds of amphibian, avian and mouse embryos.

ii) *Sonic hedgehog* is expressed in a zone of polarizing activity (ZPA) in fin and limb buds.

iii) *Sonic hedgehog* is inducible by RA in zebrafish pectoral fin, as it is in tetrapod limb buds. (RA also induces apoptosis in regenerating fins and limbs.)

iv) *Msx* is expressed in the distal limb mesenchyme in fill and limb buds.

There have been several proposals to explain the transitions between fin and limb buds and fin and limb skeletons using these expression patterns, but it may be that the genetic signals are encoded not in expression patterns, but in regulatory sequences or cascades. Tabin and Laufer proposed more distal (ectopic) activation of *Hoxa* and *Hoxd* clusters as one way to respecify distal mesenchyme in the fin→ limb transition, but more recent data does not support activation of *Hox* genes in a new distal domain. Sordino and colleagues developed a scenario from the initial expression of *Hoxd* 9 to -13 and *Hoxa*-9 to -13 in posterior and proximal limb regions, followed by progressive expression more anteriorly and distally within tetrapod limb buds, but restricted anterior expression in fin buds of *Danio rerio*.

More detailed analysis of patterns of *Hox* gene expression in avian limb buds confirm and expand Such a hypothesis. Support for differential extension of a prior proximal domain accompanied by differential regulation comes from three lines of investigation. One is detailed fate-mapping of the chick wing bud and the demonstration that cells already expressing *Hoxd*-13 'fan out' into the distal bud. The second is the cloning of 23 *Hox* genes from the embryonic chick and detailed analysis of their deployment in three phase during limb bud development. The third is targeted mutations in mice of *Hoxa-11* and *Hoxd-11* which result in loss of: the radius and ulna. Indeed, targeted deletion of *Hox* gene, in the clusters *Hox-9* to -13 is accompanied by elimination of successively more distal limb elements, indicating selective control over limb development. Eliminating *Hoxd*-13 interferes with limb size rather than limb patterning. In chick limb development, *Hoxd*-13 regulates growth at the growth-plate stage while *Hoxd*-11 acts at the prior cartilage condensation stage.

Yokouchi and colleagues misexpressed *Hox*-13 in the entire chick limb bud using a replication competent retroviral system. Misexpression of *Hoxa*-13 (like *Hoxd*-13 dramatically slowed growth and differentiation of more proximal limb cartilages (radius/ulna, tibia/ fibula). These skeletal elements displayed substantial arrest of chondrocyte growth and differentiation. This change was interpreted as a homeotic transformation of the proximal cartilages to a more distal type, producing short cartilages characteristic of the carpus or tarsus. Over-expression of other *Hox* genes does not produce this phenotypic shift. The cartilage cells in Hoxa-13-deficient embryos are more adhesive, fail to separate, and so produce smaller elements.

Even if not a homeotic transformation, this study links cell-surface responsiveness with overall skeletal patterning in an unexpected way and suggests a genetic mechanism for transformation of proximal into distal elements in the fin → limb transition. Favier et al. demonstrated effective functional cooperation between nonparalogous *Hox* gene products in developing mouse limbs and differential control over foreand hindlimbs. Targeted disruption of *Hoxa-10* had minimal effect on the forelimb skeleton, but affected proximal elements in the hindlimb. Mice made double mutant for *Hoxa-10/ Hoxd-10* showed a synergistic response with exaggerated defects in the forelimb. A further analysis, targeting mutation of *Hoxd-12* resulted in relatively minor defects in the distal forelimb skeleton: reduced length of metacarpals and phalanges and malformation of one distal carpal. Mutation of *Hoxd-13* produced a more pronounced phenotypic effect on growth and eliminated many of the metacarpals and phalanges.

Again, overlapping domains of single classes of *Hox* genes provide a potential mechanism for evolutionary transformations. Davis and Capecchi and Zakany and Duboule assessed interactions between *Hoxd-11, -12,* and -13 by interbreeding mutant strains. Interaction between these three 5' *Hox* genes was demonstrated by the presence of phenotypes not seen with mutations of individual genes. The homozygous mice resulting from this triple inactivation had small digit primordia, disorganized cartilage and impaired skeletal growth, suggesting involvement of this gene in early prechondrogenic condensations.

Similarly, targeted mutation of *Hoxa-9* and *Hoxb-9* produces more severe rib and sternal defects (including increased penetrance and increased expression of the fusions), than does mutation of either gene alone. Such studies provide further support for an interactive and synergistic role for *Hox* genes in patterning the limb skeleton and a mechanism for changes in the fin→ limb transition. Sequential spread of expression across buds in the transition from fins → limbs is likely to be sufficient to produce additional mesenchyme distally, and even to specify that the elements that develop are distal rather than proximal, but not to specify them as limb elements.

Sordino and colleagues emphasized that delayed folding of the ectoderm in the fin bud would permit prolonged signalling and mesenchymal proliferation, and allow expression boundaries of *Hoxd* genes to shift more anteriorly and distally. Such a proposal effectively combines mechanisms of altered timing and modified gene expression. An overlapping regulatory cascade of *Hoxa*-9-13 to *Hoxd*-9-13, operating along the P-D axis, may be a sufficient signalling system to explain reduction of the exoskeleton and distal extension and transformation of the endoskeleton.

Patterning of limbs: evidences of evolution

The understanding about prechondrogenic condensations are patterned into skeletal elements in tetrapod limbs increased dramatically with the work of Pere Alberch and his colleagues in the 1980s. They identified branching and segmentation as the two major processes involved in subdividing and extending condensations and patterning developing skeletal elements.

The resulting, bifurcations establish primary limb axes consisting of humerus → ulna→digital arch in the forelimb, and femur → fibula → digital arch in the hindlimb. Proximally the axis runs proximo-distally, but distally, the digital arch shifts preaxially and so runs from posterior to anterior, except in salamanders, where the digital arch extends antero-posteriorly.

Arguments based on modification of branching patterns of skeletal condensations cannot offer a sufficient explanation for the origin of tetrapod digits, for they do not address how altered branching would produce novel structures. Postaxial branching that produces digits is the branching of already established chondrogenic condensations.

Branching of proximal, condensations, followed by growth and extension of the condensations distally and anteriorly, would produce novel distal condensations as the positional shift of a developmental precursor. Such branching, growth and extension are not sufficient to produce novel structures without a second, probably *Hox*-gene-based, mechanism to transform the newly positioned proximal condensations into novel distal elements, and perhaps a third mechanism to establish joints.

This may be a situation where the fossils will help to reveal the answers, just as they revealed ancestral polydactyly and the precedence of pectoral over pelvic fins. A potential molecular basis for the subdivision of condensations into individual skeletal elements is now available with the unexpected finding by Storm and Kingsley that a BMP-like molecule named GDF-5 is expressed in transverse stripes across many developing skèletal structures of embryonic mice in positions that correspond to sites of joint formation. Furthermore, null mutations of GDF-5 disrupt formation of over 30%, of the synovial joints in the limb and change the patterns of elements of the digits, wrists and ankles. Other growth differentiation factors (GDF-6 and GDF-7) also are expressed in joint regions.

It will be of great interest to explore the distribution and any possible role of GDF-5 and other GDFs in the earlier branching and segmentation events in limbs, and to search for their expression and/or role in fin development. Perhaps nothing in vertebrate history, indeed, nothing in the history of palaeontology, is dramatic as the discovery of *Archaeopteryx* and the origin of birds, flight and feathers. This story was told in detail in 1985 in the proceedings of an international conference on Archaeopteryx edited by Hecht *et al.*, and updated in 1996 by Alan Feduccia in *The Origin and Evolution of Birds,* and again in 1997 by Bob Carroll in *Patterns and Processes of Vertebrate Evolution.*

Although palaeonto-logical evidence is revealing the origins of birds, neither palaeontological nor developmental evidence yet speaks more than a whisper about the origins of feathers. A single fossil feather was discovered in 1861.

Soon thereafter, two virtually complete fossils with feathers and wishbones were unearthed, and Huxley made the dramatic announcement that *Archaeopteryx* was precisely what Geoffroy had hoped for and Darwin's theory had predicted—an intermediate between two major vertebrate groups, reptiles and birds. Had feathers not been preserved in these specimens, without question, *Archaeopteryx* would have been classified as a reptile, for the only non-reptilian feature in the skeleton is the furcula. Huxley thought dinosaurs were the most likely reptilian ancestors of birds. His theory lay fallow for more than a century until John Ostrom disinterred it and advocated bipedal theropod dinosaurs as the ancestors of birds.

Bipedality was already established in dinosaurs by the Middle Triassic; *Sinosauropteryx prima,* from Liaoning province in northeastern China, lived some 120-140 mya and was bipedal. As might be expected from knowledge of the embryonic development of skeletal systems and epidermal appendages such as scales and feathers, and as verified from the fossil record, the transition to birds was gradual. Different features altered at different rates. A newly discovered fossil, *Unenlagia comahuensis,* intermediate between theropod dinosaurs and birds, illustrates beautifully the mix of primitive theropod and derived avian features found in such 'missing links'. Unfortunately, it is not known whether Unenlagia was feathered or possessed anything resembling protofeathers, although it was able to fold its forelimbs in a most birdlike manner. Recent discoveries of fossil birds with feathers and more specialized wings and pectoral girdles than *Archaeopteryx,* but with dinosaur-like pubic bones and ribs, indicate both how rapidly birds diversified and how they retained a mosaic of reptilian and avian characters.

Although only a little younger than *Archaeopteryx,* these Chinese and Spanish fossils had already diversified into several lineages. The transition from reptilian scales to avian feathers is not as dramatic as it might first appear, especially when we consider that common initial steps are found in the embryonic development of scales whether fish, reptilian or avian, few-hers, hairs, teeth and indeed of glands.

All arise through epithelial-mesenchymal interactions in which the initial steps promote epithelial and/or mesenchymal proliferation to facilitate folding of the epithelium and/or aggregation of mesenchyme into a placode. Indeed, epithelia from different tissues, even from different species or classes of vertebrates, can be exchanged and the initial steps in these interactions initiated, as demonstrated in interspecific recombinations between tissues from reptilian and avian embryos. Feathers or scales as alternate differentiation states, can be evoked from chick epidermis in experimental recombination between epidermis and feather- or scale-forming mesenchyme. Major issues remain in connection with this innovation, including the question of how the secondary and tertiary stages of feather development and structure (shafts, barbs, barbules) were added to the developmental programme; the parallel with the problem of the addition of tetrapod digits to a pre-existing fin skeleton will be appreciated.

Did feathers evolve for flight or for continuation of a function such as insulation already exercised by reptilian scales? Fossils remain mute on these Points, despite the vociferousness of palaeontologists. Feathered dinosaurs have not yet been found, although a 120 my old specimen housed in the Geology Museum in Beijing has a row of what appear to be appendages along the dorsal midline. Although no account of the specimen has been published in the refereed scientific literature, popular accounts attest to intense debate over whether these are appendages, and if they are, whether they are feathers, protofeathers, a skin frill as seen in some lizards, or something entirely different.

Unfortunately, development does not speak to these questions as eloquently as it speaks to us about the origin of limbs, and thus far, palaeontology is also mute on whether flight arose in tree-dwelling reptiles which used feathers to aid in gliding or in round dwellers Such as dinosaurs without a gliding intermediate. All tetrapods except turtles carry their ribs on the inside of the pectoral or pelvic girdles and limbs.

Evolution of the turtle shell (carapace) involved a shift in this fundamental arrangement. Although boid snakes such as pythons are limbless, they possess vestigial pelvic girdles and often a vestigial femur. Boids may also represent an exception to the rule of pelvic girdles outside the ribs. In turtles, each trunk vertebra corresponds to and is intimately associated with a dermal shield or osteoderm, a dermal ossification that arises in the skin independently of the endoskeleton. The carapace consists of vertebrate and ribs as endochondral elements from the axial skeleton, and fused osteoderms as dermal elements. Expansion and fusion of the ribs to one another and to the dermal ossifications in the skin produce the carapace outside the ribs.

As structure (shell), by position (outside the ribs) and by process (fusion between elements), the carapace is novel. Relationships of higher categories of turtles may be well-resolved; turtle ancestry is much less well-resolved. Ancestors have been sought in numerous groups of tetrapods, but turtles are now generally thought to have arisen from a primitive amniote group, the procolophonids, of which the Upper Permian South African genus *Owenetta* is the oldest and most completely known.

Procolophonids and testudines (turtles) form a monophyletic group on the basis of nine shared skull characters. It appears, however, that skull and shell evolved quite independently. The oldest known turtles are *Proganochelys* from the Upper Triassic of Germany, *Palaeochersis talampayensis* from the Upper Triassic of Argentina, and *Australochelys* africanus from the Lower Jurassic of South Africa. These specimens are known from skulls and, for Proganochelys, also from post-cranial skeletons. *Australochelys* and *Proganochelys* share many primitive features.

Australochelys (which is younger) is more advanced, possessing such features as a fused basipterygoid articulation, a middle ear partially enclosed laterally, and a temporal roof extending posterior to the opisthotic. *Proganochelys* is presumed to have been partially aquatic; *Australochelys* is from an arid, terrestrial environment. Fossil evidence for the evolution of turtle shells is minimal—Lee used evidence from pareiasaurs (large, herbivorous anapsid reptiles) for clues to the progression that may have also led to the origin and elaboration of turtle shells.

In a study emphasizing post-cranial modification, he argues that pareiasaurs are the nearest relatives or sister group of turtles. Some dwarf, heavily armoured forms paralleled turtles in having elaborate (but quite different) dermal armour (derived only from osteoderms) and short, fused presacral vertebrae. Lee hypothesizes that shells arose gradually by correlated progression of several organ systems.

The following sequence describes the acquisition of features described in pareiasaurs (but remember that pareiasaur shells are not turtle shells, which form from ribs as well as from dermal elements):

i) The most primitive arrangement and the earliest evidence for dermal armour in the tetrapod fossil record is the presence of small, isolated osteoderms on the dorsal midline. The postulated function of these osteoderms is not protection but postural; they provide additional area for insertions of axial muscles.

ii) Osteoderms increased in size and spread over the dorsal body surface. Although most stayed separate, some fused over the pectoral and pelvic girdles.

iii) Osteoderms fused in later forms to form solid shells.

It is unclear whether fusion of osteoderms with vertebrae and ribs in turtles required a fully fused shell of osteoderms or occurred before osteoderm fusion was complete. Whichever the case, Lee argues for extrapolation of the sequence from pareiasaurs to turtles, in part because the dermal armour in two other groups of reptiles was elaborated in the same sequence. He also uses pareiasaurs to follow the reduction in vertebral number and expansion and widening of the ribs associated with production of an inflexible body. With inflexibility came ankylosis of vertebrae and ribs and then, in turtles, fusion of ribs and vertebrae with osteoderms; or so goes the scenario.

Such gradual, small-step, correlated progression is just what is expected in organisms where organ system interact inductively during development and growth, and functionally at maturity. Lee interpreted the morphology of the anterior body of pareiasaurs as consistent with migration of the turtle shoulder girdle posteriorly into the rib cage, reversing the normal tetrapod arrangement. Burke demonstrated just such an event using developmental evidence. As determined in the snapping turtle *Chelydra serpentina*, the margin of the carapace appears as a ridge on either side of the body. This carapacial ridge consists of a mesenchymal core overlain by thickened epithelium as seen in limb buds. A unique relationship between this carapacial ridge and the rib primordia sets the stage for development of the shell. Rib primordia migrate dorsally following the expanding carapacial ridge until they terminate in the margin of the carapace.

In all other vertebrates the ribs move ventrolaterally and so remain medial to the pectoral girdles. Burke postulated a causal connection mediated by epithelial-mesenchymal interaction between the carapacial ridge and rib mesenchyme. In all other vertebrates, somitic cells that form the ribs migrate ventrally so that the ribs are positioned on the inside of the pectoral girdle. In turtles, alteration in the pathway of migration of somitic cell populations places rib-forming cells outside the pectoral girdle close to the dermis, where epithelial-mesenchymal interactions modify skeletogenesis to allow the ribs to fuse, forming the shell. The result is a novel *Bauplan.* Although the epithelial-mesenchymal interaction, as yet to be demonstrated experimentally fact that important phases of skeletal development are decoupled in the turtle *Chelydra serpentina,* demonstrates the requisite plasticity skeletogenesis.

An essentially similar scenario of inductive flexibility explains the repositioning of pouch rudiments which form internal cheek pouches in some mice and squirrels but external pouches in pocket gophers and kangaroo rats. The evolutionary origins of external cheek pouches and turtle shells are two examples of heterotopy, the topic. The ability of cichlids to adapt so extensively and rapidly in what has been called an 'evolutionary avalanche' is attributed by Liem to both prospective adaptive zones and a unique morphological key innovation involving pharyngeal jaws. This key innovation and environmental opportunity have gone hand in hand, and rapid isolation has been facilitated by behavioural traits.

The evolutionary novelty identified by Liem has three facets:

i) The development of a joint between the lower pharyngeal jaws;
ii) A shift in insertion of the fourth levator externi muscles; and
iii) The development of a joint between the tipper pharyngeal jaws and the base of the cranium.

Three such drastic changes constitute much more than a single key innovation, as indeed they do at the level of structural change. Liem, however, refers to it as a 'specialized, highly integrated innovation' and views the entire cichlid pharyngeal jaw apparatus as the key innovation. He documents plasticity in the formation of the basipharyngeal joint, incorporating knowledge gained from experimental studies in other vertebrates of the ontogeny of joints and their responsiveness to biomechanical factors. Periosteal and perichondrial cells at joint surfaces are remarkably adaptive, with the ability to modulate cellular differentiation between osteo- and chondrogenesis.

Because of the sliding of the two bony surfaces against one another, simple contact between the upper pharyngeal jaws and the cranial base would initiate the development of a cartilaginous pad between the two bones. The development of such joints is a simple and rapid ontogenetic process." Liem views these shifts as rapid and as minor genetically and developmentally. Upper and lower pharyngeal jaws and the visceral skeleton are coupled structurally, mechanically and functionally in such fishes as centrarchids (sunfishes), but decoupled in cichlids.

Liem argued that a shift of insertion of the fourth levator externus muscle facilitated by adaptive alterations in the lower pharyngeal jaws was the primary innovation.

In a recent comparative analysis of cichlid and centrarchid fishes, Galis and Drucker argue that the shift in insertion is secondary to a decoupling of the epibranchials from the upper pharyngeal jaws. The independence of movement of upper and lower pharyngeal jaws allowed greater biting force, which was reinforced by a subsequent shift in muscle insertion.

In both these scenarios the key innovative event (or sequence of events) had its origin in the developmental plasticity of the skeleto-muscular system and the ability to decouple previously coupled elements. Galis and Drucker see four ways in which such novel structures could arise:

i) Excess structural capacity;

ii) Decoupling of developmental pathways;

iii) Variability in developmental pathways; and / or

iv) Phenotypic plasticity.

Schaefer and Lauder tested for decoupling of the morphologically complex jaw articulation region of loricarioid catfishes. Their prediction that clades with decoupled systems would exhibit greater morphological variability than outgroups without decoupling, was confirmed in 10 of 12 comparisons. The two exceptions could be explained by lack of a tight association between morphological and functional evolution.

Such rigorous phylogenetic tests hold great promise for revealing the nature of the integration of morphologically complex regions Liem undertook an analysis based on comparisons of the similar development between oviparous pomacentrid and viviparous embiotocid fishes, and he and George Lauder developed the phylogenetic methodology discussed below for such studies. Meyer argues that these different morphologies represent intermediate steps in sympatric speciation.

According to this view, formulated by West-Eberhard from studies on phenotypic plasticity, developmental plasticity is an important element driving morphological change. Lauder and Liem proposed a phylogenetic methodology for testing the role of key innovations as causative agents in the initiation of morphological novelty. This methodology, the culmination of work by two of the more innovative researchers in functional morphology, builds on their studies on form and function in fishes, of which Liem's study on the origin of pharyngeal jaws in cichlid fishes is an exceptionally fine example. They provided a six-step methodology:

i) Define the key innovation or morphological novelty.

ii) Propose the consequences of possession of the innovation, i.e. propose a hypothesis to test.

iii) Map the innovation onto the group being studied and onto the appropriate outgroup.

iv) Characterize the innovation in the groups under 3, i.e. quantify the form or function associated with the innovation.

v) Perform appropriate statistical analyses on the data generated.

vi) Repeat the analysis on other groups.

Carrier developed a conceptually similar approach for the identification of evolutionary constraints and applied it to the constraints that the dual functions of running and breathing impose on hypaxial muscle architecture in some lizards. Zweers used similar approaches in his analyses of multiple roles associated with several measures of functional performance in feeding mechanics in birds. In his analysis of relationships within parasitic platyhelminths, Rohde emphasized the importance of selecting homologous characters before undertaking any analysis of key innovations and adaptive radiations. A major innovation in the successful speciation and radiation of insects in the Carboniferous was the evolution of wings. But did such innovative appendages arise from transformation of existing structures or *de novo* as novelties?

Two major (and many minor) hypotheses for their origin have been proposed: wings either developed by modification of limbs or as new structures independent of the limbs. Both scenarios are considered briefly, although the jury is still out on which (or which other) scenario best explains the origin of wings. According to the modified limb hypothesis, wings evolved from a segment of the legs of wingless (apterygote) insects. Given that limbs are ventral and wings are dorsal appendages, origination from limbs involves extensive migration around the body. Unfortunately, little more can be said about the likely changes between limbs and wings. However, from fossil evidence we know that the first winged (pterygote) insects had multiple pairs of wings, perhaps involving all thoracic and abdominal segments.

Insects, which have one or two pairs confined to the thoracic segments, could have evolved from this primitive condition by reduction in numbers of pairs of wings. Given what is known of how homeobox genes control segment identity in insects, and given that the primitive condition was one of repeated pairs of wings along the body, modern insects may have come by their reduced number of wing pairs by suppression of homeobox genes, a possibility investigated by Sean Carroll and his colleagues in 1995. Their novel results involved a close investigation of *Antennapedia* in a winged insect (Drosophila) and the wingless thysanuran *Thermobia domestica*.

Wings and halteres (the modified second pair of appendages in Drosophila that transform into wings in such mutants as *Bithorax*,) develop on the second and third thoracic segments. Although development of the segments is dependent on *Antp*, development of wings and halteres is not. Other homeotic genes control wing initiation and patterning. Carroll and colleagues propose that wings arose without input from homeotic genes. Subsequent modification of Scr led to the elimination of wings on the prothorax, while evolution of Ubx or abd-A eliminated wings on insect abdominal segments and on the mesothorax of dipterans.

The latter possibility is strengthened by the recent demonstration that ectopic expression of engrailed in Drosophila transforms the anterior, but not the posterior, compartment of halteres to wings through negative regulation of Ubx by *engrailed*. Homeotic genes such as Ubx do control segment identity. Gibson and Hogness identified polymorphisms of Ubx in Drosophila, with phenotypic effects on thoracic segment three. Their strategy was to use artificial selection for differential sensitivity to the induction of phenocopies of the *Bithorax* phenotype using ether vapour as the environmental trigger.

Increased sensitivity to ether correlated with loss of expression of Ubx in the imaginal disc for the third thoracic segment. There is also potential for regulation through modification of the 120kb regulatory region of Ubx. There are several variants of the second proposal, which follows from the hypothesis that wings did not evolve from segments of the legs. In one, wings are novel appendages unrelated to any pre-existing structures, limb or otherwise. In another, proposed by William Carpenter in the 19th century, wings evolved from gills.

Averof and Cohen investigated this proposal by searching for genes in a crustacean, the brine shrimp, Artemia *francoscana,* that serve functions specific to wing development in winged insects. They found two genes—*pdm,* which is thought to be involved in initiation of wing primordia in *Drosophila,* and apterous, which determines DV patterning of *Drosophila* wings. Both are associated with the 'respiratory limb' in *Artemia-pdm* at specification of the limb buds, apterous with a dorsal expression pattern. Unfortunately for the 'wings from gills scenario' other genes involved in patterning appendages are not shared between *Artemia* and Drosophila.

The approach, based on shared patterns of expression, assumes conservation of function of these genes between wingless and winged organisms. This is no more parsimonious than the assumption that genes in a common ancestor were independently modified in crustaceans and insects for roles in appendage development and is an excellent example of the difficulty of using either patterns of gene expression or shared developmental pathways as the basis for determination of homologies. A key innovation need not be structural; it could be genetic, cellular, developmental, physiological, behavioural or ecological.

As an example of a genetic change, a mutation has been proposed as the key innovation in the origin of directionality of torsion in gastropods. Normally, single gene mutations of large effect are lethal. It is therefore of interest to examine the arguments marshalled for a mutation in a single gene resulting in the production of something as fundamental as directionality of early development and body organs. The shells of snails, indeed the whole of the viscera, undergo torsion or spiralling during development. Torsion is but one of the many rearrangements of organs found in molluscs, characterized by Ponder and Lindberg as visceral rotations, folds and flops. Torsion may be right-handed (dextral) or left-handed (sinistral), is species-specific, and can be traced back to the plane of the first cleavage division of the egg. Gastropods, in common with annelids, turbellarians, nematodes and all molluscs other than cephalopods, display a spiral pattern of cleavage. If the mitotic spindle in the first cleavage division angles to the right, a right-handed spiral is imposed on embryonic and adult structures. If the mitotic spindle angles to the left, a left-handed spiral is imposed.

The mitotic spindle rotates from right to left (or *vice versa*) with each successive cell division. Strictly, this is not spiral but alternating cleavage, for spiral cleavage would have a constant direction of rotation of the spindles. The cleavage pattern is more analogous to standing on the North Pole and alternately turning to the right and the left, than to spinning like a top on top of the earth.

Directionality of torsion is controlled by a single pair of genes through cytoplasmic factors inherited in the egg an instance of the maternal cytoplasmic control. This was revealed when cytoplasm from the eggs of snails with right-handed coiling was injected into eggs of left-handed coiling snails, transforming them into righthanded coiled individuals. Direction of torsion is under simple genetic control with the potential to produce an innovation of large effect. The key appears to be control of a fundamental and early developmental process by a single pair of genes, and the consequences of that process for subsequent body organization.

Environment's effects on evolution

Nineteenth-century embryologists were much concerned with adaptations for embryonic and larval existence. By and large, 20th-century developmental biologists have not dealt with adaptations and adaptability. An exception, Conrad Waddington, considered three kinds of adaptation as process:

i) *Exogenous adaptation*, in which the organism is modified to be better fitted to its special environmental circumstances. According to Baur and to Waddington, truly exogenous adaptation is selection for the capacity to respond developmentally to the environment in an appropriate way;

ii) *Pseudoexogenous adaptation*, in which an organism exhibits characteristics similar to those of exogenous adaptation, but the characters are hereditary and independent of any particular environmental stimulus. Waddington believed that the genome imposes limits on the degree to which the phenotype could respond to environmental change; and

iii) *Endogenous adaptation* or functional connections, as of the cornea to vision.

Population geneticists emphasize the adaptive nature of the relationship of populations to their environment. American palaeontologist Edwin Colbert endeavoured to show that adaptation could be especially visualized in series of fossils in which progressive adaptation over long periods of time could be related to environmental change. Few fossil lines show progressive adaptation, and when it is absent, the problem of how to infer environmental change arises.

Striking exceptions are the studies of bovid speciation and extinctions in sub-Saharan Africa undertaken by Vrba, and similar studies of hominids by Brain. Both these workers correlate speciation and extinction events that parallel adaptive shifts with global temperature variations that influence rainfall and vegetation. Low temperatures at the end of the Miocene epoch five mya, and again 2.6 mya, were associated with greater rates of speciation of hominids and antelopes than at intervening times when temperatures were rising. Antelope distribution ranges also fluctuated in concert with the climatic and vegetational changes.

To study adaptation from the standpoint of functional morphology, we need to find an approach that will include an environmental component and supplement the mathematical models of the population geneticist.

One measure, morphological disparity or morphological distance, has been used in two studies of quite disparate organisms and structures: molar tooth diversity in Caenozoic ungulates and morphological change in post-Palaeozoic crinoids. These analyses concentrate on morphological change independent of taxonomic status to reveal ecological components driving morphological change or morphological change that appears to transcend ecological changes. Jernvall and colleagues developed an innovative approach to adaptation in their study of molar tooth diversity in Caenozoic ungulates in relation to ecological and climatic changes.

Their use of morphological distance follows the methodology introduced by Raup. Tooth crown types rather than taxonomic units form the basis of their analysis of adaptive radiations during the 50 million years of the Caenozoic. The response of different taxa which share particular adaptive radiations to such global changes as climatic cooling could be analysed using similar approaches. Although it has long been thought that slow temperature change facilitates evolution, the notion that abrupt alteration in climate may facilitate periods of rapid evolution is novel.

Even deep-sea organisms, such as benthic ostracods, exhibit fluctuations in diversity over time scales of 10000 to 100000 years that correlate with glaciation, diversity decreasing as glaciers advance and recovering during interglacial periods. Climate also influences evolutionary change over shorter periods. An example is change in cranio-dental characters in the pocket gopher, *Thomomys talpoides,* that track climatic change over 3200 years of the Upper Holocene. Coope discovered similar correlations for Caenozoic beetles. Although the morphology of beetle species was remarkably stable during the enormous climatic oscillations of the Quaternary period, their geographical ranges changed enormously: the beetles tracked the tolerable environment across continents, in the process breaking down geographical barriers between populations to allow genetic mixing and prevent speciation.

In cases like this, there are obvious difficulties in identifying adaptive responses. The adaptation may be reflected in morphology, speciation, or alteration of geographical range. As the latter is a physiological and behavioural adaptation, only large-scale study of fossil faunas over considerable geographical areas will reveal it, or allow it to be inferred.

The reigning view is that adaptation to the environment occurs by the natural selection of mutations whose direction of change is entirely random with respect to the environment, but which, by chance, give rise to individuals with suitable characteristics. In phenotypic plasticity, which is one class of evolutionary plasticity, when the environment is heterogeneous, a single genotype can develop different phenotypes in different environments'.

Phenotypic plasticity modulates the genetic response of organisms to selection. In terms of heritability, it is the proportion of total phenotypic variance accounted for by genotype by environment interactions. Epigenetic control directs the building of organisms through inductive interactions (embryonic induction) in which signal and response occur within the same organism. But there are other classes of inductions, including situations, such as cyclomorphosis, in which the products of one organism's genes activate specific developmental programmes in another organism of the same or different species, as in interactions between a predator and embryos of the prey species. There are further situations where the inductive signal originates from a dietary constituent from a food source, or from abiotic sources in the environment such as light and temperature.

In other situations, induction by another species or from the environment can trigger the same response, either because of independent induction under different circumstances or through cooperative induction. Therefore, the following classes of induction can be identified:

i) Induction between cells within an individual embryo, larva or adult. This is embryonic (interorganismal) induction.

ii) Induction between individuals of the same species, as in pheromone-based signalling in insects and density-dependent polymorphism in salamanders. This is intraorganismal (intraspecific) induction.

iii) Induction between individuals of different species, often between predator and embryos of a prey species. This is interspecific induction, with predator-based induction as a subclass.

iv) Induction initiated by abiotic, environmental signals such as light and temperature. This is environmental induction.

v) Inductions involving combinations of an environmental signal and any of the other four classes of induction.

These types of induction appear to be quite distinct, but, all share the ability to activate a developmental pathway in embryos of a responding species that leads to the production of distinctive embryonic, larval or adult morphologies. Fundamentally, they differ only in the source of the inductive signal. In many instances cyclomorphosis arises in response to inductive stimuli from outside the responding organism, usually produced by a predator.

A chemical released by a predator (or obtained from a food source in environmental induction; evokes a morphological change in the offspring of the prey species (or species consuming the food), switching development into new morphogenetic and differentiative pathways. The genetic component is the ability of a prey species to respond by selective gene expression to activate a new developmental pro gramme. Consequently, alternate or multiple phenotypes can be derived from one genotype in response to different environmental, ecological or inductive signals. This is the ecological equivalent of the phenotype of the gene introduced in section 7.4. Schmalhausen formalized such responses as norms of reaction, a concept now virtually synonymous with phenotypic plasticity. The stimulus that evokes the specific developmental pathway is environmental as far as the responding species is concerned, but involves the production and release of a gene product as far as the predator is concerned. Such interspecific, predator-based responses are not odd rarities. They occur in plants, algae, protozoans, ciliates, rotifers, cladocerans, bryozoans, gorgonians, gastropods, barnacles, insects, frogs, amphibians, reptiles and fish.

Interspecific interactions are epigenetic because of the specific response of the prey species; the interaction is epigenetic at the level of developmental response. Other examples, in which a developmental programme is elicited by different signals, include phenocopies and the relationship between genetic assimilation and homeotic mutations. Seasonal polymorphism provides some of the most compelling evidence for the relevance to organismal ecology of inductive control of development. Perhaps even more importantly, it links evolution, ecology, environment and community-level interactions to individual development.

Cyclomorphosis enables organisms to avoid predation, take advantage of seasonal food supplies, compensate for overcrowding, and tolerate and/or avoid extreme environmental variation. Cyclomorphosis is a dramatic illustration of Leigh Van Valen's aphorism that evolution is the control of development by ecology. The ecological importance of interactions is clear when we realize that *B. calyciflorus* individuals with the additional set of spines cannot be eaten by the predator. Gilbert subsequently showed that different quantities of prey elicit different *Asplanchna* morphs and that efficiency of feeding varies from morph to morph. A second rotifer species, *Keratella testudo*, responds to soluble factors released by some ten competitive zooplankton species, including species from other phyla. Additional spines are produced, body size is larger, and survivorship higher, all clear indications of the ecological advantages of cyclomorphosis.

Even more impressive as an indication of the developmental potential and responsiveness within such species is the finding that single clones of *K. testudo* produce spines of different types and sizes in response to signals from different predators. A suite of morphological responses is available, providing a remarkably adaptive plasticity for species that encounter many predators at different seasons. Similarly, copepods that undergo extensive daily vertical migration within the zooplankton and for which vulnerability to predation varies with migration and life-cycle stage, show different ontogenetic patterns adapted to different combinations of predators encountered. From such studies we see that, not only do such interspecific inductions integrate the life histories of predator and prey, they integrate communities of organisms.

An important but little-studied area of evolutionary biology is the origin, development, maintenance and ecological significance of such relationships between predators and prey. Sharon Emerson, Harry Greene and Eric Charnov made a beginning in 1994 when they advocated the application of scaling models used in ecology and functional morphology. The production of 'neckteeth' in progeny of a cladoceran, *Daphnia pulex*, that is preyed upon by larvae of the phantom midge, *Chaoborus*, parallels the rotifer example, including the release of a diffusible molecule by the predator.

Abiotic environmental agents can also evoke alterations in helmet formation; 10 hours exposure to 5μg/l of the carbamate pesticide carbaryl alters helmet form at the second instar. Seasonal changes in head and tail spines and helmet morphology in various species of *Daphnia* do not involve any change in genotypic frequency—they are genuine polymorphisms. The altered *Daphnia* have ecological consequences themselves, for they induce changes in their own food supply, a green alga, which grows too large to be consumed by the *Daphnia*. These essentially developmental interactions are important elements of the trophic ecology of the species and community. One vertebrate example is the development under natural or experimental conditions of a deep-bodied morph of the crucian carp, *Carassius carassius*, in the presence of the predatory pike *Esox lucius*.

Divergence in body shape is seen as early as 12 weeks after introduction of the predator. The faster rate of growth and increased drag that the deeper body entails are offset by the final size of the deep-bodied morphs, which are too large to be eaten by the pike. As with the invertebrate examples, polymorphism provides a refuge from predation. Adaptation to predation is not unique to species exhibiting polymorphism. Various aspects of life history (body size, allotment of resources to reproduction, fecundity, size of offspring and so forth) respond to predator pressure because predators are size-specific in their selection of prey. Thus, guppies (*Poicilia reticulata*) show significant changes in life history parameters in as little as 30 generations of response to altered predation. Polymorphism is a specialized form of such life-history adaptation. An elegant demonstration of the inductive nature of associations between species is seen in interaction between the marine luminescent bacterium *Vibrio fischeri* and the squid *Euprymna scolopes*. These squid are luminescent because of the light emitted by symbiotic bacteria housed within epithelial crypts of their light organs. The bacteria invade the developing juvenile light organ through ciliated cells. The symbiont engineers its own entry; ciliated cells die only in the bacteria's presence. Furthermore, epithelial cells of the light organ only increase in size to produce the crypts if bacteria are present

In the absence of bacterial 'infection' the light organs remain in a state of arrested morphogenesis, an elegant example of specific inductive interaction between species. The entomology literature contains many examples of regulation of development through interactions between individuals of the same species. Such interactions are very highly developed in the social insects (ants, bees, social wasps) with different castes (queen, soldiers, workers) within a single population. One example that especially appeals to me is regulation of the production of soldier ants in *Pheidole bicarinata*. The number of soldiers produced is regulated by a combination of interactions between individuals and the colony's nutritional status. Adult soldiers in this species release a pheromone which inhibits production of further soldiers by regulating the threshold of larvae to juvenile hormone, the hormone that triggers development of soldiers. Size and allometric relationships of soldier ants are regulated through programming imaginal-disc growth patterns. Hormone levels are sensitive to nutritional status of the colony: a balance of nutritional status and numbers of soldiers determines whether additional soldiers will form.

The development of larval bees into workers or queen depends on whether or not they are fed the 'royal jelly' secreted by the worker's salivary glands—an example where nutrition alone determines phenotype. Some amphibians are polymorphic with respect to whether they metamorphose or remain as permanent aquatic adults. Less well-known are the cannibalistic morphs of two subspecies of the tiger salamander (*Ambystoma tigrinum nebulosum* and *A. t. mavortium*) and of the New Mexico spade-foot toad, *Scaphiopus multiplicatus*.

Diet and overcrowding respectively trigger development of large, aggressive individuals whose morphology and behaviour enables them to cannibalize their smaller siblings. Cannibalistic morphs have enlarged heads, much enlarged jaws and feeding apparatuses (including altered teeth and shortened intestines) enabling them to feed on other amphibian larvae. *Scaphiopus* feeds on crustaceans with high thyroid hormone content, and the cannibalistic morph can be induced by the application of thyroid hormone. The changes are not atypical of metamorphosis, just as the changes in the ants were mediated by juvenile hormone.

In terms of a classification of epigenetic factors that can be incorporated into quantitative genetics models of the evolution of development, a distinction needs to be made between situations where the signal is a gene product and situations where it is an environmental stimulus. Rotifers are an example of the former, light-eliciting chlorophyll production in plants of the latter. Such clean distinctions between biotic and abiotic environmental signals are not, however, always possible; for example, predator and environment often interact to elicit the response. A conceptually similar situation is the artificial activation of eggs by a temperature or pH shock in the absence of sperm; the stimulus may be abiotic, the response is not. The distinction between biotic or abiotic signalling is not critical at the level of analysis of the activation of developmental pathways within a single generation.

It is important in the context of models of the heritability and evolutionary stability of such interactions. A fine example of environmental induction is revealed by Erick Greene's study of the geometrid moth *Nemoria arizonaria*, which made the cover of *Science.*Caterpillars of this North American and Mexican moth hatch out on several species of oak trees in either spring or summer. Those that hatch in spring feed on oak catkins low in tannin; those that hatch in summer, after the catkins have fallen, feed on oak leaves that have a high tannin content. Caterpillars are morphologically similar when they hatch but become strikingly different as they develop; they become seasonal polymorphs. Spring caterpillars mimic catkins and are therefore referred to as catkin morphs; summer caterpillars mimic oak twigs and are known as twig morphs.

Catkin morphs have bright yellow skin with many papillae and processes and two rows of stamen-like dots along the midline; twig morphs have grey skin and fewer processes. Greene eliminated temperature and photoperiod as causal factors in morph production and demonstrated that the amount of tannin in the diet controlled morph type. Catkin morphs developed from 94% of larvae fed catkins, but from only 6% of larvae fed leaves.

Ecological correlates of this remarkable seasonal polymorphism allow caterpillars to overcome the apparent disadvantage of feeding on such a seasonal food supply as oak catkins. Catkin morphs pupate more rapidly, have higher survival to pupation and are larger at pupation than are twig morphs. Furthermore, females that metamorphose from catkin morphs have greater fecundity than do those that metamorphose from twig morphs. Greene emphasized what may be an unrecognized widespread occurrence of diet-induced developmental polymorphism in the evolution of host specificity and host races, confirming the alternative adaptations theory of incipient speciation put forward by West-Eberhard.

West-Eberhard, who hypothesized that the mechanisms in such interactions play a major role in speciation, postulated that the maintenance of different adaptive phenotypes in the same life history stage within a population is the first step to speciation. Having established alternate morphs within a species, reproductive isolation allows speciation from one of the morphs.

Given that polymorphisms are developmentally based and induced, her approach places epigenetic control at the base of phylogenetic bifurcation, linking developmental mechanisms directly to speciation.The genetic basis of environmentally induced responses is now under investigation, especially in plants, with the search for 'plasticity genes', which Pigliucci defines as regulatory loci that directly respond to a specific environmental stimulus by triggering a specific series of morphogenic changes'. Alterations in rates of DNA transcription and of RNA translation are known in at least four systems displaying phenotypic plasticity: the response of flowering plants to light, induction of heat-shock proteins in plants and animals, adaptive responses of cyanobacteria to sulphur limitation, and temperature-dependent sex determination in lizards.Environmental and biotic signals can independently elicit the same morphological responses in *Daphnia* (induction of neekteeth) and salamanders (development of cannibalistic morphs).

A predator-based induction and an environmental signal can also cooperate to elicit the response, as demonstrated by Michael Bell and his colleagues in an elegant analysis of reduction in the pelvic girdle of the threespine stickleback, *Gasterosteus aculeatus*, in 179 lakes surrounding Cook Inlet, Alaska. The pelvic girdles are comprised of pelvic spines, posterior and anterior processes, and an ascending branch. The pelvic spines can be elevated and locked upright and so deter predators.

Physiologically, burst speed and escape from predation is correlated with other skeletal characters, such as length of vertebrae, vertebral number, and the ratio of abdominal to tail vertebrae. Pelvic girdle reduction occurs in response to absence of predation by other fishes and low levels of dissolved calcium in the lakes. It occurs in stages. To go is the pelvic spine, then the posterior process, the ascending branch, and finally the anterior process.

Phylogenic classification and evolotion

Cladistics is a rational method of inferring evolutionary relationships and portraying them in a logical classification, typically a dichotomously branching tree, or cladogram. For a classification to meaningfully reflect evolutionary history, the organisms comprising each group (e.g., species, genus, family, phylum) should be defined by a unique characteristic, or set of characteristics, called a homologue. For example, gill slits, notochord, dorsal hollow nerve cord, and postanal tail are considered by most to be homologues that uniquely define a group of animals called chordates. All chordates express these homologues, at least at some point in their life history, and in all other animals this particular set of homologues is absent. The urochordates, cephalochordates, and vertebrates are assumed therefore, to have diverged from a common ancestor and are more closely related to, each other than any one is to another group of animals, that is, they are monophyletic. If, instead, each subphylum had evolved independently from a separate ancestor, the chordates would be polyphyletic in origin.

Variations in the structure of homologues between the subphyla provide additional information useful for inferring the evolutionary relationships of the chordates. Foscusing on the notochord, recall that it is expressed as a coelom in the urochordates (larvaceans), as a stack of muscle cells in the cephalochordates, and as a cord of noncontractile, fibrous cells in the vertebrates, all variations on a common theme. Because a coelom occurs in phyla other than the chordates and because the cephalochordate notochord develops enterocoelously, it therefore seems likely that the coelomic layout of the urochordate notochord is the common theme. It is a carryover from the coelomate ancestor of the chordates, and as such is regarded as the primitive, or plesiomorphic, form of the notochord. The muscular notochord of the cephalochordates and the fibrous notochord of the vertebrates (both with well-developed sheaths) are variations on the theme and are considered to be evolutionarily derived, or apomorphic, forms of the notochord.

Just as a common theme is general and its variations specific, so too a primitive expression of a homologue characterizes a broad, more inclusive group, whereas derived expressions of a homologue characterize narrow, more exclusive, groups. As a result, a classification based on homology identification is multileveled, or hierarchic, in structure. Because homologous similarities are believed to be the result of evolutionary descent with modification, such a classification is viewed as an evolutionary (phylogenetic) classification.

After considering such an obvious example, it may be surprising to learn that few of the major evolutionary lines of metazoans are as well defined as that of the chordates, and many groups of phyla, for example the lophophorates or the aschelminths, lack even one uniquely distinguishing homologue. For this reason, the interrelationships of most of these phyla and their specific placement within the general framework of metazoan phylogeny is uncertain.

Phylum-level relationships suggested by the application of electron microscopy are:

i) Metazoans are probably a monophyletic kingdom, based on the widespread occurrence of choanocyte-like cells and uniflagellated sperm.

ii) Metazoans probably evolved from colonies of flagellates that resembled the extant choanoflagellates, based on similarities of choanoflagellate and metazoan flagella.

iii) Coelomates probably constitute a monophyletic taxon, because of similarities of their mesothelia, blood vessels and nephridia.

iv) Protostomes and deuterostomes are each likely to be monophyletic taxa, based on the nature of their photoreceptoral cells.

v) Nemerteans are probably coelomate animals, as indicated by the peritoneal lining of the rhynchocoel and "blood vessels," as well as the schizocoelous mode of development of the -blood vessels."

vi) Developmental and structural evidence supports a close relationship between sipunculans and molluscs.

vii) Gastrotrichs and nematodes (possibly nematomorphs) constitute a monophyletic group, based on similarities of their pharynx and cuticle.

viii) Rotifers and acanthocephalans probably constitute a monophyletic group because of the similarity of their integument.

ix) Pentastomids and crustaceans may be a monophyletic group based on similarities of their sperm.

x) Pogonophorans, echiurans, and annelids may constitute a monophyletic group because of the identical structure of their setae.

General evolutionary trends and reinterpretations documented by electron microscopy are:

i) Cells bearing one flagellum (or cilium) are more primitive than those bearing more than one, a conclusion supported by comparative and developmental analyses.

ii) Ciliary photoreceptors are probably more primitive than microvillar (rhabdomeric) photoreceptors, based on comparative and developmental studies. Combined with item 4 in the previous list, this inference suggests that the deuterostomes are more primitive than the protostomes.

iii) An epithelial musculature is probably more primitive than a subperitoneal Musculature in coelomates, based on comparative and developmental Studies.

iv) Cross-striated muscle and a type of smooth muscle, in which the thin actin filaments attach and anchor to the cell membrane only (e.g., in cnidarians and ctenophores), are more primitive than other types of striated and smooth muscle arrangements.

v) Neurons having multiple functions (sensory-motor interneurons, as in some cnidarians) are more primitive than those having a specialized function.

vi) A sperm composed of an acrosome, spherical nucleus, short mitochondrial midpiece and a single flagellum is more primitive than other designs of sperm.

vii) Filtration nephridia are the primitive excretory organs of bilaterians.

viii) Coelomates having blood vessels and metanephridial systems can evolve into acoelomates having protonephridia and lacking blood vessels, as evidenced in many species of annelids. Such "regressive" evolution is actually a positive adaptation to small body size.

ix) Many small nematodes, all gastrotrichs, and kinorhynchs are acoelomate and not pseudocoelomate in body organization.

x) There is evidence of a true coelom, lined by peritoneum, in a species of the pseudocoelomate Priapulida.

xi) Species that have a cuticle can evolve into species that lack one, as evidenced in the, Annelida. As with the loss of a coelom, the loss of cuticle is correlated with small body size or the adoption of an endoparasitic habit.

Phylum-level evolutionary indications from molecular analyses are:

i) Although the ciliates constitute a monophyletic group, the flagellates are almost certainly polyphyletic in origin (rRNA).

ii) The Metazoa is probably a monophyletic taxon (rRNA).

iii) Rhombozoan "mesozoans" are probably rooted in the diploblastic animals and not in the bilaterians (rRNA).

iv) Coelomates are probably a monophyletic group (rRNA).

v) Protostomes and deuterostomes are each likely to be monophyletic taxa (rRNA).

vi) Platyhelminths are an early, perhaps the earliest, lineage within the Bilateria (rRNA).

vii) Nemerteans and coelomates are likely to form a monophyletic taxon (rRNA).

viii) Sipunculans and molluscs are probably a monophyletic group (rRNA),

ix) Annelids and pogonophorans are likely to be a monophyletic taxon (rRNA).

x) Arthropods are probably a monophyletic group (rRNA, hemocyanin).

xi) Within the coelomate protostomes, the annelids and molluscs may form one monophyletic taxon, and the arthropods another. If this is true, the traditional closer relationship between annelids and arthropods would be rejected.

The ultrastructural data have more or less supported the possibility that an acoelomate body can evolve from a coelomate design, thus supporting Remane's evolutionary viewpoint. Hyman's position, however, is supported by preliminary molecular data. They suggest that the acoelomate platyhelminths evolved before the lineage, or lineages, leading to the coelomate animals. At this point in our discussion, it may be useful to construct a series of phylogenies of metazoans to illustrate better the provisional nature of evolutionary classifications in general and to highlight hypotheses of general significance, in particular the evolution of bilaterally symmetrical animals.

The approach of evolutionary classification of metazoan can be based on three homologues: eukaryotic cells, tissues, and organs, which are often regarded as levels of organization. The assumption here is that the eukaryotic cell evolved in the protists and was retained, not independently evolved, in the lines to the fungi, metaphytes, and metazoans. The eukaryotic cell, therefore is a homologue of protists, fungi, metaphytes, and metazoans and is one indication that they form a monophyletic taxon. Similarly, once tissues, such as epithelial and connective tissues evolved in lower metazoans, they were retained in higher metazoans. Thus, animal tissues are unique to metazoans, they define metazoans as monophyletic, and they distinguish metazoans from protozoans. The same is assumed to be true for the capacity to organize tissues into organs.

Although sponges and placozoans have tissues, they lack the organs found in all other metazoans and can be distinguished from them on that basis. The capacity to form organs, therefore, is a homologue of "higher" organ-bearing animals setting them apart as a monophyletic taxon within the Metazoa as a whole. Thus, these three homologues provide a first approximation to the phylogeny of metazoans.

Sponges and placozoans are largely asymmetric in body form; cnidarians, ctenophores, and "mesozoans" are radial or biradial, as are the early developmental stages of most other animals; but bilateral symmetry distinguishes adult higher animals from cnidarians and ctenophores. At the next level of consideration, degrees of compartmentation of the body by epithelia can be used as characteristics with which to further specify the evolutionary tree. Although a protoepithelium occurs already in sponges, a true epithelial epidermis is expressed, in placozoans, an epidermis and gastrodermis (with connective tissue in between) occur in cnidarians and all higher animals. A subset 4 the higher animals, called coelomates (e.g., annelids, arthropods, molluscs, echinoderms) have a third epithelium (mesothelium) that encloses a coelom, a homologue unique to the coelomate assemblage of phyla.

The occurrence of a mesothelium splits the Bilateria into the primitive acoelomates and pseudocoelomates, and the evolutionarily derived coelomates. Although bilateral symmetry unambiguously defines the bilaterally symmetrical animals, alternative classifications result from assumptions regarding the primitive or derived states of other characteristics. In the earliest bilaterians, did the blastopore give rise to the mouth or anus, or to the mouth and anus? Were they coelomate or acoelomate? Was the nephridium a protonephridium or a metanephridial system? Was a blood-vascular system present or absent? Were the animals large or small? Depending on how these questions are answered, different evolutionary classifications result. The unlikely assumptions are that some presumed homologue evolved more than once. Spiral cleavage, however, is diphyletic, having evolved once in the acoelomates and again in the coelomates. Ultimately, a choice between these alternatives, or the establishment of another more accurate tree, lies in the discovery of new facts and a careful analysis of homology.

Acoelomate organization implies small body size, a characteristic that may be primitive in Bilateria, but one that also has evolved independently in several groups of animals. The advent of small body size in aquatic animals is correlated with the use of cilia for locomotion, the absence of a coelom and blood-vascular system, and the occurrence of protonephridia, all of which are homologues of acoelomates. Because all of these can be reduced to a single trait, small body size, the assemblage may be polyphyletic rather than monophyletic in origin. With this possibility in mind, the spiralian acoelomates may be said to include the gnathostomulids, platyhelminths, and entoprocts. The gnathostomulids are regarded as the primitive taxon in the assemblage because their epidermal cells and protonephridial terminal cells bear only one cilium each. In the entoprocts and platyhelminths, on the other hand, the epidermal and protonephridial cells are multiciliated, a shared-derived homologue.

One lineage, set apart by the advent of segmentation, embraces the annelids, pogonophorans, echiurans, onychophorans, and arthropods. Although the echiurans lack segmentation, they are grouped with the annelids because of the shared occurrence of an annelidan cross. The absence of segmentation in echiurans, therefore, is viewed as secondary feature. The second lineage includes the sipunculans and molluscs, which are shown as having evolved from a nonsegmented ancestor with few coelomic cavities (oligomery). An alternative to this phylogeny is one in which segmentation appears earlier, in the ancestor to both annelids and molluscs. In that case, the absence of segmentation in sipunculans and molluscs, except perhaps in Neopilina, would have to be interpreted as a secondary loss.

The deuterostomes, including the chaetognaths and lophophorates, may constitute a monophyletic taxon based on the common occurrence of enterocoely and a tricoelomate body, but the specific placement in the tree of phyla in relation to each other is problematic. Placement of the lophophorates in the deuterostomes suggests that the absence of enterocoely in bryozoans and phoronids is a derived characteristic, as is the derivation of the mouth from the blastopore, a typical protostome trait, in the phoronids.

If it is assumed that hemichordate and chordate gill slits are homologous structures and that the complex protocoelic nephridia of hemichordate and echinoderm larvae are also homologues, then the absence of gill slits in echinoderms is probably secondary. This scenario is supported by preliminary molecular data from the analysis of rRNA sequences. An alternative is to consider the hemichordates and chordates together as a monophyletic group and the echinoderms as a separate, more primitive taxon within the deuterostomes. In that case, the homologue, gill slits, would uniquely characterize the hemichordates and chordates.

The protocoelic nephridum, however, would either have had to evolve twice independently, once in the echinoderms and again in the hemichordates, or it evolved in the common ancestor to all three phyla. If so, the chordates have secondarily lost the protocoelic nephridium. The systematic position of the chaetognaths relative to other deuterostomes, a perennial question in deuterostome phylogenetics, remains unanswered for lack of a shared-derived homologue that links them with any other phylum.

The aschelminth assemblage of phyla has long been placed between acoelomates and coelomates as a group of evolutionarily intermediate animals. Although this may be a correct interpretation, a significant amount of new information speaks against a close relationship (monophyly) of these taxa and even rejects the idea that all are pseudocoelomate. As a result, it seems likely that some aschelminth phyla, or groups of them, could be more closely related to some non-ascheliminth phylum, then they are to each other. The question is: to which non-aschelminth phylum or phyla is any aschelminth group linked? Rotifers; and acanthocephalans provide evidence of spiral cleavage, but does this indicate an alliance with the acoelomate protostomes or the coelomate protostomes?

The probable monophyly of gastrotrichs, nematodes, and nematomorphs has already been mentioned.

All have bilateral cleavage patterns, and in many gastrotrichs, the blastopore forms both the mouth and anus. Should these three phyla be aligned with protostomes or deuterostomes? The unfortunate fact is that on the basis of current information, it is impossible to find a characteristic that uniquely links any aschelminth phylum with any non-aschelminth phylum. A thorough phylogenetic analysis of a group of organisms as large and diverse as the Metazoa involves such complex issues and such a magnitude of information as to be beyond the comprehension of any one individual.

44

12/1033/44
11/5/02